NATURAL
SELECTION and HEREDITY

HARPER TORCHBOOKS / The Cloister Library

(*continued on next page*)

HARPER TORCHBOOKS / The Science Library

(continued on next page)

P. M. Sheppard   NATURAL SELECTION AND HEREDITY. Illus. TB/528
O. G. Sutton   MATHEMATICS IN ACTION. Foreword by James R. Newman. Illus. TB/518
Stephen Toulmin   THE PHILOSOPHY OF SCIENCE: *An Introduction*   TB/513
A. G. Van Melsen   FROM ATOMOS TO ATOM: *The History of the Concept* Atom   TB/517
Friedrich Waismann   INTRODUCTION TO MATHEMATICAL THINKING. Foreword by Karl Menger   TB/511
W. H. Watson   ON UNDERSTANDING PHYSICS: *An Analysis of the Philosophy of Physics.* Intro. by Ernest Nagel   TB/507
G. J. Whitrow   THE STRUCTURE AND EVOLUTION OF THE UNIVERSE: *An Introduction to Cosmology.* Illus.   TB/504
Edmund Whittaker   HISTORY OF THE THEORIES OF AETHER AND ELECTRICITY: *Vol. I, The Classical Theories,* TB/531; *Vol. II, The Modern Theories,* TB/532
A. Wolf   A HISTORY OF SCIENCE, TECHNOLOGY AND PHILOSOPHY IN THE 16TH AND 17TH CENTURIES. Illus. *Vol. I,* TB/508; *Vol. II,* TB/509

HARPER TORCHBOOKS / The Academy Library

James Baird   ISHMAEL: *A Study of the Symbolic Mode in Primitivism*   TB/1023
Henri Bergson   TIME AND FREE WILL: *An Essay on the Immediate Data of Consciousness*   TB/1021
H. J. Blackham   SIX EXISTENTIALIST THINKERS: *Kierkegaard, Jaspers, Nietzsche, Marcel, Heidegger, Sartre*   TB/1002
Walter Bromberg   THE MIND OF MAN: *A History of Psychotherapy and Psychoanalysis*   TB/1003
Abraham Cahan   THE RISE OF DAVID LEVINSKY: A Novel. Intro. by John Higham   TB/1028
Helen Cam   ENGLAND BEFORE ELIZABETH   TB/1026
G. G. Coulton   MEDIEVAL VILLAGE, MANOR, AND MONASTERY   TB/1020
Wilfrid Desan   THE TRAGIC FINALE: *An Essay on the Philosophy of Jean-Paul Sartre*   TB/1030
John N. Figgis   POLITICAL THOUGHT FROM GERSON TO GROTIUS: 1414–1625: *Seven Studies.* Intro. by Garrett Mattingly   TB/1032
Editors of *Fortune*   AMERICA IN THE SIXTIES: *The Economy and the Society*   TB/1015
G. P. Gooch   ENGLISH DEMOCRATIC IDEAS IN THE SEVENTEENTH CENTURY   TB/1006   *
Francis J. Grund   ARISTOCRACY IN AMERICA: *A Study of Jacksonian Democracy*   TB/1001
W. K. C. Guthrie   THE GREEK PHILOSOPHERS: *From Thales to Aristotle*   TB/1008
Henry James   THE PRINCESS CASAMASSIMA: A Novel   TB/1005
Henry James   RODERICK HUDSON: A Novel. Intro. by Leon Edel   TB/1016
Henry James   THE TRAGIC MUSE: A Novel. Intro. by Leon Edel   TB/1017
Arnold Kettle   AN INTRODUCTION TO THE ENGLISH NOVEL: *Vol. I, Defoe to George Eliot,* TB/1011; *Vol. II, Henry James to the Present,* TB/1012
L. S. B. Leakey   ADAM'S ANCESTORS: *The Evolution of Man and His Culture.* Illus.   TB/1019
Bernard Lewis   THE ARABS IN HISTORY   TB/1029
Arthur O. Lovejoy   THE GREAT CHAIN OF BEING: *A Study of the History of an Idea*   TB/1009
Niccolo Machiavelli   HISTORY OF FLORENCE AND OF THE AFFAIRS OF ITALY: *From the Earliest Times to the Death of Lorenzo the Magnificent.* Intro. by Felix Gilbert   TB/1027
J. P. Mayer   ALEXIS DE TOCQUEVILLE: *A Biographical Study in Political Science*   TB/1014
John U. Nef   CULTURAL FOUNDATIONS OF INDUSTRIAL CIVILIZATION   TB/1024
Robert Payne   HUBRIS: *A Study of Pride*: Foreword by Herbert Read   TB/1031
Samuel Pepys   THE DIARY OF SAMUEL PEPYS: Selections, ed. by O. F. Morshead; illus. by Ernest H. Shepard   TB/1007
Georges Poulet   STUDIES IN HUMAN TIME   TB/1004
Priscilla Robertson   REVOLUTIONS OF 1848: *A Social History*   TB/1025
Ferdinand Schevill   THE MEDICI. Illus.   TB/1010
Bruno Snell   THE DISCOVERY OF THE MIND: *The Greek Origins of European Thought*   TB/1018
W. H. Walsh   PHILOSOPHY OF HISTORY: *An Introduction*   TB/1020
W. Lloyd Warner   SOCIAL CLASS IN AMERICA: *The Evaluation of Status*   TB/1013
Alfred N. Whitehead   PROCESS AND REALITY: *An Essay in Cosmology*   TB/1033

NATURAL SELECTION and HEREDITY

P. M. SHEPPARD

Senior Lecturer in Genetics
in the University of Liverpool

HARPER TORCHBOOKS / The Science Library

HARPER & BROTHERS, NEW YORK

NATURAL SELECTION AND HEREDITY

Printed in the United States of America

First published in the Hutchinson University
Library, in the Biological Sciences Division,
edited by H. Munro Fox. Reprinted by arrangement with Hutchinson & Co., Ltd., London. The
original edition was published in 1958 and a
revised edition in 1959.

First HARPER TORCHBOOK edition published 1960

CONTENTS

PREFACE

The Darwin–Wallace papers read to the Linnean Society of London on 1st July, 1858, and Darwin's book *On the origin of species*, published in 1859, mark a turning point in biology. Not only did they present evidence that evolution had occurred, but they explained it in terms of natural selection.

It is necessary that the characteristics of the organism be inherited if selection is to result in change; but no progress was made in understanding the laws of heredity until 1900, when Mendel's work, published in 1866, was rediscovered.

Thus genetics (the study of heredity and variation) is no older than the present century, but it has been the subject of a vast amount of work. Progress in analysing natural selection and evolution was slow at first, since this had to await the development of genetic principles, but it has been rapid during the last thirty years.

This book deals with some recent ideas on the mechanism of evolution in the light of modern genetics. Because of the importance of genetics in evolutionary theory, two chapters explaining the elementary principles of heredity have been included, in order to help those with no previous knowledge of this subject.

The amount of information on natural selection and its mode of action is so great that it would be impossible to treat it fully in ten volumes the size of this book. Consequently I have selected certain subjects which illustrate important principles and have treated them in detail. I believe that this approach will lead to a better understanding of the subject than an elementary survey of the whole field of population genetics, which, for lack of space, must be wholly inadequate. In consequence of this I have had to leave out many subjects, such as the evolution of sex-determining mechanisms, selection in bacteria, hybrid vigour (particularly important in the development of hybrid corn), the

7

nature of the gene and human evolution, to mention only a few.

The most important topic I have omitted, perhaps unwisely, is the evolution of the chromosomal system of heredity itself, on which depends the orderly distribution of the genes and, consequently, the course of subsequent evolution. I have not specifically covered this subject, other than discussing the effects of polyploidy and inversions, because it has been brilliantly covered by C. D. Darlington in his book *The evolution of genetic systems* and because space does not allow its adequate treatment here.

While writing this book I have been influenced by three considerations: (i) to explain principles, (ii) to avoid giving the impression that most problems concerned with natural selection are solved and that controversy no longer exists, an idea which elementary books often impart to their readers, (iii) to reach a standard sufficiently advanced to allow those who wish to follow up the ideas in this book to do so by reading original papers.

In order to achieve these objectives I have reduced the number of examples to the minimum and have tended to pick well-known groups of animals to illustrate my points. Consequently the Lepidoptera (butterflies and moths) are well represented, for this group is as well known to most people as any, with the possible exception of the birds. Moreover, they have been extensively used in evolutionary studies and I myself have worked with them on occasion. I have included in the text some hypotheses which are supplementary to, or alternatives to, those that are generally accepted, thus emphasizing the incompleteness of our knowledge a hundred years after the Darwin–Wallace lecture. I have not hesitated to use scientific terms where this leads to brevity, as familiarity with them is necessary for reading, without difficulty, more advanced books and papers. Space does not allow me to give a glossary but the technical terms are explained in the text. Nor can I give full references for all my statements, but I have often given the author to whom I am referring. The actual reference can usually be found from the books and papers listed at the back of this book.

I am most grateful to Dr. A. J. Cain who drew the illustrations and found time to read the manuscript in detail. His many comments have been of the greatest use to me. Dr. E. B. Ford,

F.R.S., has also spent much time and trouble in reading the manuscript, and I am deeply indebted to him for his encouragement and constructive criticisms. Without the help of Dr. H. B. D. Kettlewell and his kindness in allowing me to use unpublished data the account of industrial melanism would not have been nearly so detailed or accurate. I also wish to acknowledge the help and advice I have received from Dr. L. P. Brower, Dr. C. A. Clarke, Sir Julian Huxley, F.R.S., Professor K. Mather, F.R.S., Professor R. J. Pumphrey, F.R.S., Dr. S. Walker and Dr. M. Williamson. I am entirely responsible for the choice of subjects and for any omissions or errors which may occur. However, without the help of so many colleagues and friends the text would have been less comprehensible and several errors might have been included. I also wish to thank Miss B. Sanderson for the care with which she typed the manuscript and the editors of *Genetics* for allowing me to reproduce Figure 6.

P. M. SHEPPARD

Department of Zoology
Liverpool University

I

NATURAL SELECTION

JULY 1st 1858 marks a turning point in biological thought, the immense importance of which could hardly have been appreciated at the time. It was on this date that the Darwin-Wallace lecture was delivered before the Linnean Society of London. Subsequently it was published in the third volume of the Society's Journal.

Both Charles Darwin and A. R. Wallace independently put forward the view that species (p. 181) are not individually created and unchanging, but that each could give rise gradually to new species during the course of time. That species are not immutable but can change, or *evolve* as we would now say, was not a new view. However, the point which was new (it had been put forward before in a tentative way but with little detailed argument and evidence to back it) was the hypothesis that natural selection is the essential agent directing and controlling such change.

Before discussing the hypothesis in more detail, it is desirable to mention briefly some of the beliefs held by biologists at that time. Broadly speaking, there were two schools of thought on the origin of the various forms that have inhabited, or still inhabit, the earth. One school maintained that species are individually created and are unchanging. It was agreed by many that slight deviations from the normal form occurred from time to time. However, it was also held that in the end these variations always reverted to the original form, and could not be sufficiently distinct to constitute new species. The other group, which included such men as Erasmus Darwin (Charles Darwin's grandfather), and the great French biologist Lamarck, held the view that species could change gradually into new species, in other words that they could evolve.

Lamarck's great contribution to biological thought was to support the theory of evolution with cogent arguments. He also put forward an hypothesis as to the factors controlling evolutionary change. He maintained that living matter had an inherent capacity to alter gradually over many generations from a simple structure or organization to a more complex and perfect one. Over and above this, he noted that organs which are much used tend to become larger and more highly developed as the result of this use, compared with those in an individual in which they are not so extensively exercised. Moreover, he observed that when they are not used at all, they tend to diminish in size. He assumed that such modifications, acquired by an organism during its own lifetime, may be inherited to a certain extent by the offspring. Consequently he accounted for the remarkably delicate and complex structure of many organs, which is so well suited to their particular function, by modification during the course of generations as the result of inheritance of these 'acquired characters'. For example, he postulated that water birds, in their efforts to swim, extended their toes and so stretched the skin between them. The stretched condition would be inherited, he thought, and the process repeated in the offspring of the birds, and in subsequent generations, until a webbed foot had been evolved. Thus species would gradually become well suited, that is *adapted*, to their environment, and once they had done this they would remain constant in structure until conditions changed. He supported his hypothesis by many such examples and, particularly, by pointing out that domestic animals and plants, during their domestication, had departed from their wild ancestors in a remarkable way.

It is greatly to Lamarck's credit that he made out such a strong case for evolution at a time when many of his colleagues believed firmly in individual acts of creation to account for species. It is not surprising that he believed in the inheritance of acquired characters, for this was a reasonable hypothesis to propose at that time. Moreover, the mechanism of inheritance was probably a consideration of secondary importance to him as compared with the task of convincing people that evolution occurred.

It seems probable that Lamarck's insistence that species

evolved gradually, a feature essential to his hypothesis, was in part due to his own observations and deductions, and in part to the fact that the sudden appearance of a new species would be too strongly suggestive of special creation. Charles Darwin certainly agreed with Lamarck in the belief that species evolved slowly and, in part at least, for the same reasons. He also followed Lamarck in believing in the inheritance of at least some acquired characters, and it seems not unlikely that Lamarck's work impressed Darwin with the contribution that the study of domestication could make to knowledge about evolution, thus stimulating his own interest in it.

The Darwin-Wallace lecture consists of four papers. The first is a letter from Sir Charles Lyell and Dr. J. D. Hooker stating the circumstances leading up to the reading of both Darwin's and Wallace's papers at the same meeting. They explained that in 1839 Darwin put down his views in a manuscript, which, according to them, was rewritten in 1844 and shown to Hooker. Wallace sent his own paper to Darwin in 1858 with the request that, if it was of any interest, it should be sent to Lyell. Consequently Lyell and Hooker deemed it advisable to put forward both views, which had been independently arrived at, at the same meeting of the Society. Darwin's contribution consists of part of his manuscript together with a letter sent to Professor Asa Gray of Boston in 1857.

In these two papers he gives the reasons which led him to make the suggestion that natural selection was a primary factor in controlling the course of evolution. He had been much impressed with the theoretical work of Malthus on the growth of human populations. He pointed out that animals, like man, produce sufficient offspring to ensure that the population will increase at an enormous rate if all survive. To illustrate this, Darwin gave the following example: 'Suppose in a certain spot there are eight pairs of birds, and that *only* four pairs of them annually (including double hatches) rear only four young, and that these go on rearing their young at the same rate, then at the end of seven years (a short life, excluding violent deaths, for any bird) there will be 2048 birds instead of the original sixteen.' He then pointed out that, as Malthus had realized, in fact populations do not increase in this way but remain on the average fairly constant in size. However, they have the potentiality of increasing their numbers at a great rate, as is demonstrated by the fact that populations of plants and animals under favourable conditions, especially when introduced into a new country, sometimes expand very rapidly, although a stable state is reached in time. He concluded that the numbers of organisms in each country must be kept in check 'by recurrent struggles against other species or against external nature'. It is quite clear that he does not necessarily mean struggle in the physical sense, but simply to express the idea that conditions, or the *environment*, as we now call them, are always harsh, and get worse with increasing population size. Under such conditions there is competition both between species and between individuals of the same species for space, food, protection from enemies, the acquisition of mates and similar commodities in short supply.

Having reached this point, Darwin explains his hypothesis in such a clear way that I cannot do better than quote him. 'Now can it be doubted, from the struggle each individual has to obtain subsistence, that any minute variation in structure, habits or instincts, adapting that individual better to the new conditions, would tell upon its vigour and health? In the struggle it would have a better *chance* of surviving; and those of its offspring which inherited the variation, be it ever so slight, would have a better *chance*. Yearly more are bred than can survive; the smallest gain in the balance, in the long run, must tell on which death must fall, and which shall survive. Let this work of selection on the one hand, and death on the other, go on for a thousand generations. Who will pretend to affirm that it would produce no effect, when we remember what, in a few years, Bakewell effected in cattle, and Western in sheep, by this identical principle of selection?'

Darwin believed that in any environment an organism will accumulate in the course of time the *inheritable* variations which best fit or adapt it to its surroundings. If the environment changes, new variations will become advantageous, will be utilized and will supplant the old less well-adapted forms. Darwin realized that for this 'natural selection' to be effective the variations must be inherited, and that there will have to be a store of them present at the moment when they can be utilized. Noting that variation often appeared when environmental conditions were altered, as in the domestication of wild species, he postulated that a change in conditions brings about the spontaneous production of new variability (now known as mutation, p. 51). We know now that this is not true (if we except the results of an increase in ionizing radiation and the presence of certain chemicals), but that the new variation appears as the result of a change in the direction and intensity of the selection itself. Thus, as is explained in Chapter VI, the change in selection 'releases' inherited variability which, although present before, did not exert a visible effect. Darwin concluded from his arguments that advantageous variations, however small, would be preserved, and that as the result of their accumulation the species would depart further and further from its original form.

Other aspects of selection and evolution more fully dealt with in his later works, including his great book *On the origin of species* published in 1859, were mentioned in his lecture. One in particular must be considered here. Darwin separated sexual selection from other types. He pointed out that there exist in animals which have two sexes, particularly in the male, many attributes which seem unlikely to contribute to the survival of the individual and may even be deleterious to it. He easily explains the presence of these. For example, if a male possesses a structure or behaviour-pattern which stimulates the female in such a way that his chance of securing her for a mate in the presence of a rival is increased, the attribute will put him at an advantage with respect to the number of progeny he leaves. Moreover, such a character will be improved and perfected in the course of time because any variations increasing its stimulating power will put their possessor at an advantage, and these more successful males will leave more offspring to future generations compared with the less successful ones. The character will only cease to change when its sexual advantage is exactly counterbalanced by some mechanical, physiological, or other disadvantage—that is to say, when sexual selection, as Darwin calls it, is counterbalanced by equal and opposite natural selection. In this way he explained the otherwise inexplicable development of many secondary sexual characters, such as the plumes of Birds of Paradise or the peacock's tail.

Darwin clearly over-emphasized the importance of sexual selection, for he included too wide a category of phenomena under this heading. Many, such as the songs of birds, are as much concerned with the intimidation of other males and the maintenance of territory as with the attraction of a female. However, it is equally obvious that some recent authors have under-estimated its importance.

Wallace's paper approached the problem of evolution from the same point of view as Darwin's, but with rather different emphasis. He pointed out that those who thought that species were unchanging entities admitted that variation occurred. However, they held that this variation could not be large enough to constitute a new species, and that new forms always reverted in time to the original species type. Wallace pointed out that little was known about varieties in nature, and that the argument was based on information from domestic forms. He maintained that such generalizations could not be supported by the consideration of domestic species, and that there was no justification in applying information from them to variation found in the state of nature. He then went to considerable lengths to emphasize, as Darwin did, that organisms have the capacity to increase their numbers very rapidly, but that in fact they do not, their numbers tending to remain fairly constant over considerable periods of time. Moreover, the common species are not necessarily the ones which have the greatest fecundity. Consequently he argued that population size depended not on the potential reproductive capacity of the species, but on other factors, particularly food supply.

Wallace was undoubtedly correct when he maintained that in general factors other than reproductive potential control population size. Food shortage may often be such a factor,[48] but Wallace clearly put too much emphasis on this particular item. Modern ecologists know of many others, for example interactions between the density of predator and prey, and host and parasite, to mention only two (see p. 174). Wallace suggested that if we knew enough about such factors, we would be able to determine why one species was common and another closely allied form was rare. It is a sobering thought that 100 years later in not a single instance are all the controlling factors for a wild population known and, in fact, hardly one such factor is fully understood.

As the result of his argument, Wallace put forward two propositions:

(i) An animal population generally remains, on the average, constant in size, its potential increase being checked by food shortage and other factors.

(ii) The comparative abundance or scarcity of species is entirely due to their efficiency in combating these controlling factors.

Wallace further maintained that any variation from the normal form would affect the capacities of the organism. He suggested that this effect will render the individual either more efficient, or less efficient, in combating the controlling factors. Thus a variety with a reduced capacity to obtain food, shelter, etc., would in the end perish. However, those better in these respects would increase in numbers, and it would be the original form, and not the variety, which would perish.

Now if conditions changed in a particular area, the species would be unlikely to be as efficient in combating the environment, and would perish, but if a variety was present which was less severely affected, it might survive and replace the old form of the species. He also admitted that there might be some organs which could vary and not affect the capacities of the species, and he suggested that variations in them might persist together with the parent form. Darwin came to much the same conclusion. He said 'Variations neither useful nor injurious would not be affected by natural selection, and would be left either a fluctuating element, as perhaps we see in certain polymorphic species, or would ultimately become fixed, owing to the nature of the organism and the nature of the conditions.' This passage is of immense interest. Darwin was wrong in believing that the variation in polymorphic species was not affected by selection, as we shall see in Chapters IV and V, although it should be noted that he was cautious about the matter. In fact it is only by the action of natural selection that two or more distinct varieties can be maintained indefinitely in a population. However, if by the word 'fixed' he meant that one form would eventually replace all others, he was correct in believing that if natural selection was not operating, varieties could become fixed under certain conditions. This is particularly true when the population size is small, as is explained in Chapter VII.

Towards the end of his contribution, Wallace made one of the most telling points in his argument against those who opposed the theory of evolution by natural selection. Wallace pointed out that domesticated plants and animals have essential commodities, such as food and protection, provided by man. Therefore many of the characters necessary for survival in the wild will disappear under domestication, whereas others, useful to man but deleterious under natural conditions, will be accumulated. Thus neither a draught horse nor a racehorse is suitable for existence in a wild state. He argued from this that domesticated species, when turned loose, would either die out or return to a form close to their ancestors. This selection for an ancestral type at once explains why varieties found under domestication would not persist indefinitely, and thus why one cannot argue from domestic species for the transitory nature of varieties which occur in wild species.

It is interesting to note that because nothing was known about the mechanism of inheritance, Wallace had confused two different phenomena when talking about the permanence of varieties. He was quite correct in pointing out that varieties close to the ancestral form will survive, whereas others will be eliminated. However, many of the inherited characters found in domesticated breeds are what is called recessive (p. 36). Such characters may not be found in the progeny of an individual possessing them when it is mated to one possessing a contrasting character, as will be explained in Chapter II. Such varieties seem to disappear although the offspring and some of their descendants will carry the capacity to give rise to the 'lost' character, and as the result of an appropriate mating it will reappear. He therefore confused the elimination of unfavourable varieties by selection with the temporary disappearance of recessive varieties.

Wallace concluded by suggesting that there was no necessity

for postulating that the capacity to vary had any particular limit, as did the believers in special creation. Consequently the process he described would ensure that species would gradually change and give rise to new ones. Both Darwin and Wallace emphasized the struggle for existence due to the inherent capacity of populations to increase in size. It is a great pity that they both used the word 'struggle', which suggests physical combat, for this has led to confusion, and has obscured the all-embracing nature of selection. Evolution can occur even when there is no difference between forms in their viability (capacities to survive). For example, if there are two inherited varieties which do not differ in this respect, one would replace the other if it were more fertile, other things being equal. In other words, the fate of a form (that is to say, a variety or sport) will depend on whether the variation is inherited and on the contribution the form makes to future generations (not necessarily only the next generation). To take an extreme case, a variety which is very resistant to cold, drought, starvation and other killing agents, will not become commoner if it is sterile, however advantageous it may be in other respects. The action of selection depends on the net effect of all the different selective pressures acting, and these are seldom known in detail. Consequently the only practical way of measuring the net selection pressure is by the change in the relative proportions of different varieties from generation to generation. In practice, this is usually measured as a percentage advantage or disadvantage of one form over another.

Both Darwin and Wallace concluded that forms which are particularly well adapted to the environment are evolved by the action of selection on the individual. Darwin makes this quite clear in *On the origin of species*, when he writes: 'Natural selection acts only by the preservation and accumulation of small inherited modifications, each profitable to the preserved being. . . .' As we have seen, this is not strictly true, for an increase in fertility could not really be considered as advantageous to the individual possessing the quality, although such a capacity will be selected for if it does not bring with it some equal or greater disadvantage in another direction.

Space does not allow a review of all Darwin's beliefs and conclusions, but anyone reading his works will soon discover that he held many views as advanced as, or even more advanced than, those held by some present-day evolutionists. He was clearly far ahead of his contemporaries, and when he was wrong it will often be discovered that his mistakes spring from an absence of knowledge of the mechanism of inheritance.

Darwin was initially more concerned with demonstrating the reality of evolution, and the necessity of postulating natural selection to explain it, than with proving the existence of selection experimentally. He was apparently either not impressed with the necessity of demonstrating its reality by experiment, or, more likely, thought that in each generation its effect was so slight that its action could not be detected in the space of a human lifetime.

Dr. E. B. Ford reports that Charles Darwin, in conversation with his son Leonard Darwin, 'said that if data were properly collected, they might reveal "perhaps in no more than fifty years" the progress of evolutionary change'. The facts of the matter are that most of Darwin's examples of the operation of selection are taken from domestic animals and plants, and nowhere does he give adequate data which prove beyond all reasonable doubt the action of selection in the wild. Those of his contemporaries who supported him seemed even less concerned in demonstrating selection than was Darwin himself. Some did more harm than good to this theory by speculating wildly about the advantage of this or that character, without due consideration, and doubtless often without seeing the living animal or plant in the wild state. As a consequence of this unbridled speculation, many biologists by the beginning of the twentieth century had rejected Darwin's views. It was, in fact, not until R. A. Fisher and J. B. S. Haldane used the mathematical approach to refute the objections to Darwin's theory that the pendulum swung again towards the selectionist view.

As late as 1936 G. C. Robson and O. W. Richards[55] were able to say of examples of selection : 'It will be seen that on this analysis (which should be checked by reference to the actual accounts) there is a little evidence suggesting a significant difference between survivors and eliminated. It must be admitted that any amount of *positive* evidence, however slight, is of value. On the other hand, it is of greatest importance that, in all cases in which selective elimination appears to be established, the distinguishing features of the survivors are not known to be heritable.' It is true that in their account they are inclined to use special pleading to reduce the conclusiveness of some results they discuss. Nevertheless, that they are able to make this statement seventy-eight years after the Darwin-Wallace lecture is, in itself, remarkable.

One is inclined to think of selection tending to alter a species, rather than keep it constant. In point of fact, selection is probably usually maintaining the constancy of a character, rather than altering it. H. C. Bumpus,[5] in 1898, reported the results of measuring a number of characters in a group of house sparrows which had been incapacitated by a snowstorm in America. He found that the measurements of those that survived were different from those that did not, and concluded that this demonstrated the action of natural selection. Robson and Richards tried to throw doubt on the matter by saying 'but in reply we must, obviously, ask how any structural character (such as weight, wing-spread, etc.) which might determine whether a bird was blown down or not, could determine whether a bird survived or died after it was blown down—a result which might be determined by such purely accidental causes as whether it hit a branch or stone in its fall, or whether it was able to withstand exposure and shock'. But this is not the point. All we need concern ourselves with here is whether there were real differences between those that died and those that survived, and not the physical factors which resulted in life or death. Which sparrows died and which did not was not determined by chance alone, as is shown by the real difference in measurements between the two groups.

W. F. R. Weldon[72] in 1901 and A. P. di Cesnola[11] in 1907 demonstrated the action of natural selection on the shells of the land snails *Marpessa laminata* (Montagu) and *Arianta arbustorum* (L.). Both these writers showed that there was greater variability in shape among young snails than among adults. It was the extreme variants, departing far from the average shape, which were eliminated. In other words, natural selection was tending to stop shape from varying indiscriminately. It is true that we do not know for certain if the characters investigated by these workers and Bumpus are inherited, so that we cannot be sure that in these examples selection is effective in reducing variability. This does not, however, invalidate the demonstration of the selection itself.

We are on firmer ground when considering two other examples which have been recently investigated, for in both of these we know that the character is in part determined by inherited factors although also much affected by the environment. In one, the hatching success of ducks' eggs of various sizes was studied. As a result it was possible to show that fewer of those that were either larger or smaller than the average succeeded in hatching. In other words, here again selection was tending to maintain the optimum size. The other investigation concerned human birth weight.[29, 43] In one hospital the weight at birth was noted, as was the proportion of infants surviving for the first twenty-eight days of life, in the different weight classes. It was shown that those that were well under 8 lb. in weight had a poorer expectation of life than those near it. What is more surprising at first sight is that those above this weight also had a poorer expectation of life. Consequently, here again selection is tending to eliminate those that are too heavy or too light at birth. This type of selection has been called stabilizing, for it does not tend to alter a character, but to maintain it near the optimum for the conditions in which the individual is living. Although when Robson and Richards wrote, in 1936, there were not many cast-iron demonstrations of the action of natural selection, the number has increased rapidly since then, and there are now too many to review here, although a small proportion of them will be dealt with in later chapters.

We can now turn to some of the objections which have been put forward against Darwin's theory. These can, in general, be divided into the following categories:

(i) Difficulties felt by Darwin himself and springing from his acceptance of the hypothesis of blending inheritance (*see below*). As is noted later (p. 109) Mendelian inheritance alone allows of the retention of great inheritable variability and of keeping the species constant at one and the same time.

(ii) Special examples in which it is difficult to imagine what advantage the intermediate stages in the evolution of an organ can have for their possessor.

(iii) Misapprehension by the early Mendelian geneticists of the bearing of Mendel's discovery of particulate inheritance on the theory of evolution.

Taking these points separately:

(i) Darwin believed, as did most of his contemporaries, that inheritance is blending. It was thought that a 'substance' which determined the transmission of an inherited character was passed by each of the parents to their offspring, in which the two contributions become inseparably mixed. Consequently the offspring of a cross tended to be intermediate between the parents in characters by which they differed, and, because the inherited substance from both parents blended, the parental characters could not reappear in unaltered form in future generations. Thus if, for example, there are equal numbers of black and white cattle in a herd, under this hypothesis it would not take many generations before all were grey in appearance. This would not happen in one generation for some black animals would mate with others like themselves and give only black offspring, and similarly some whites would mate with white. However, the number of greys would increase rapidly, for a mating between white and black would produce grey offspring. Under such a system, about half the variability would be lost in every generation. Consequently, if selection is to be effective, it would have to act with enormous speed.

To overcome this difficulty, Darwin suggested that changed conditions produced new variability; but in fact this does not help, for it would then be the nature of this continuous flow of new variability which would determine the direction of evolution and not selection. For not only will all characters available to selection be of very recent origin (others having been lost by blending inheritance) but also if the rate of production of a deleterious character by spontaneous change is greater than the rate of its elimination by selection it will spread despite its disadvantage. That is to say, the direction of the spontaneous changes will determine evolution. It is surprising that so few of

Darwin's opponents seemed to appreciate these points, for they would be fatal to his theory. It was not until Mendel showed that inheritance is not blending that the difficulty could be resolved.

(ii) Objections to the theory of evolution by natural selection which rely on the impossibility of imagining advantageous intermediate stages in the development of a character can be illustrated by the electric organs of fish. These structures are capable in some fish of producing a discharge able to stun or kill quite large animals. Darwin said on the matter: 'The electric organs of fishes offer another case of special difficulty; for it is impossible to conceive by what steps these wondrous organs have been produced. But this is not surprising for we do not even know of what use they are. In the Gymnotus and Torpedo they no doubt serve as powerful means of defence, and perhaps for securing prey; yet in the Ray, as observed by Matteucci, an analogous organ in the tail manifests but little electricity, even when the animal is greatly irritated; so little, that it can hardly be of any use for the above purposes. Moreover, in the Ray, besides the organ just referred to, there is, as Dr. R. M'Donnell has shown, another organ near the head, not known to be electrical, but which appears to be the real homologue of the electric battery in the Torpedo.' Thus the difficulty felt by Darwin, and much emphasized by his less critical critics, seems to be in imagining what use the production of a few volts, not capable of causing a severe shock, can be to a fish. This difficulty is partially dispelled largely by the work of H. W. Lissmann,[49] who showed that in the African fish *Gymnarchus niloticus* Cuv. there is a constant stream of pulses emitted by the fish when it is at rest. Lissmann demonstrated that the fish is able to detect the presence of objects such as food or other fishes by means of this, in the same way as in radar. Mr. D. Moorhouse informs me that the East African fish *Mormyrus kannume* Forskal, the Elephant Snout Fish, can also detect food and other fish in a similar manner. However, this species alters the rate of emission of impulses; this is very low when the fish is at rest but increases when it is disturbed. The potential difference developed is small and amounts to about six volts. In order to detect the presence of disturbances in its own electrical field the fish has, besides the electric organ, a special receptor organ at the base of the dorsal fin. It is interesting to note that Darwin makes particular reference to a probably similar organ in the Ray. This new evidence on the advantage of an electric organ not capable of administering powerful shocks underlines the fact that it is illegitimate to criticize the theory of natural selection on grounds which involve the inability of the critic to visualize advantageous intermediate stages.

Rather similar arguments have been raised with respect to the acquisition of organs of 'extreme perfection'. For example, Darwin says: 'It has been objected that in order to modify the eye and still preserve it as a perfect instrument, many changes would have to be effected simultaneously, which, it is assumed, could not be done through natural selection; but as I have attempted to show in my work on the variation of domestic animals, it is not necessary to suppose that the modifications were all simultaneous, if they were extremely slight and gradual.' Darwin is here clearly pointing out that intermediate stages could have an advantage, and therefore the eye could gradually evolve, and it is not necessary for the organ to appear in a perfect state for it to be useful. Fisher has pointed out that far from negating the theory, evolutionists could not explain the production of such a complex organ except by postulating natural selection, for the probability of all the necessary modifications appearing initially in one organism (except by special creation) is virtually zero. It is only their gradual accumulation, as the result of natural selection, which increases sufficiently the probability of their appearing in the same individual for highly integrated systems to be evolved. In fact selection makes it probable that at least some organs of immense complexity will be evolved.

(iii) The last group of objections were mainly raised by the early Mendelian geneticists. They believed that Mendel's discovery, that inheritance is not blending but is controlled by discrete factors which do not mix with and contaminate one another, was fatal to Darwin's theory. Thus H. de Vries, W. Bateson and others followed Mivart in believing that species and higher categories appear as the result of sudden changes, or *mutations* as they would now be called. However, as already stated in the case of the eye, the production of complex integrated systems at one step is almost impossible, and arguments of this nature make the theory of the production of very different species by such mutations equally unlikely. Consequently, although it was formerly believed that conspicuous mutants or sports are important in evolution, few people still hold that view. But as the matter has been raised again recently by R. B. Goldschmidt,[30] the subject will be discussed in more detail in Chapter XII.

Those interested in arguments about the necessity for believing in evolution, and, in particular, evolution by natural selection, should read Darwin's *On the origin of species*, and an excellent chapter by R. A. Fisher in *Evolution as a process*, in which he demolishes most arguments against the theory of natural selection.

SIMPLE MENDELIAN INHERITANCE

As HAS been pointed out in the previous chapter, Darwin realized that a clear understanding of the laws of inheritance was essential in any comprehensive theory of evolution. However, at the time when *On the origin of species* was first published nothing was known about inheritance, and it was generally believed that some substance from each parent formed an inseparable mixture in the offspring and that this determined their inherited characters. Consequently, it was believed that a grandchild had a mixture consisting of one-quarter of this substance or 'blood' from each grandparent.

Such a system would be expected to produce offspring which are intermediate in all characters between the parents. It seems surprising to us, with our present knowledge, that such a belief could have been held when it must have been obvious to all who bred animals and plants that this intermediate appearance of the offspring often does not occur. In fact Darwin himself knew that this principle was not universal and observed that in the primrose there are two distinct types of flower (pin and thrum, Fig. 7, a and b). He noted that intermediates do not occur or are very rare, and that this morphological difference is the basis of outbreeding in the species (p. 93).

A belief in the commonness of blending inheritance probably accounts for the ready acceptance by Darwin of the hypothesis that characters acquired during the life of the individual, because of environmental influences, can be inherited. This arises from the fact that it is quite easy to believe that the inherited substance, if it existed, since it could be changed by being amalgamated with other inherited substances, could be changed also by environmental factors, then diffuse to the reproductive organs and be passed on in its altered state to the progeny.

27

Evolution as a result of natural selection is quite incompatible with blending inheritance, as Darwin realized (p. 23). This difficulty together with numerous exceptions to the supposed rule of blending inheritance accounts for the modified hypothesis which he put forward. In this he postulated that particles or gemmules from different parts of the body came together in the sperm and ova, and that the presence of these accounted for inheritance. Realizing that blending inheritance was not universal, he explained its occasional absence by suggesting that only some of the gemmules in a cross between different forms mixed to form a hybrid gemmule, which determined the characters of the offspring. Moreover, he thought that some did not mix but remained pure and dormant, and that in later generations some of these uncontaminated particles developed to produce again characters found in the grandparents and in earlier generations without blending. He also maintained that crossing facilitated the development of such dormant bodies.

That he was aware of many of the results of particulate inheritance is clearly shown by the fact that he said: 'Crossed forms are generally at first nearly intermediate in character between their two parents; but in the next generation the offspring generally revert to one or both of their grandparents, and occasionally to more remote ancestors.' Furthermore, in another place he says: 'When grey and white mice are paired, the young are not piebald nor of an intermediate tint, but are pure white or of the ordinary grey colour.' He was aware of many other such examples, both from other people's work and his own breeding experiments. He himself crossed, in both directions, the common snapdragon (*Antirrhinum*) with a variety having abnormal flowers called the peloric form. The ninety offspring which he examined were all of the normal type, none being pelorics. He also took the precaution of showing that plants with the peloric flowers, when self-pollinated, produced only peloric snapdragons. Describing the rest of his experiment he said: 'The crossed plants, which perfectly resembled the common snapdragon, were allowed to sow themselves, and out of a hundred and twenty-seven seedlings, eighty-eight proved to be common snapdragons, two were in an intermediate condition between the peloric and normal state, and thirty-seven were perfectly peloric, having reverted to the structure of their one grandparent.' It will be noted that the proportion of normal to abnormal flowered plants is about 3 : 1, a ratio well known in Mendelian inheritance (p. 36). In fact this experiment is similar to those carried out by Mendel on peas.

It is surprising and unfortunate that he did not deduce the true laws of inheritance from his knowledge of actual breeding data. His two main mistakes were (i) postulating that particles were formed in different parts of the body and were then transported to other places, including the reproductive organs. This was probably in part due to his belief that characters acquired during the life of an organism, as the result of the action of the environment, could sometimes be inherited. The inheritance of such acquired characters would necessitate the existence of some substance which could diffuse from the changed organs to the reproductive bodies. (ii) He believed that the particles from different parents sometimes formed an 'amalgam' and gave inherited properties intermediate between both parents. This was probably due to the fact that, though realizing that not all inheritance was blending, he believed that quite a considerable proportion of it was.

It was Gregor Mendel who laid the foundations of the modern theory of inheritance, or *genetics*, as it is now named. He gave a paper on the results of his breeding experiments to the Brünn Natural History Society in 1865 and his communication was published in the Transactions of the Society in 1866. However, the paper failed to attract attention until 1900 when C. Correns, H. de Vries and E. von Tschermak, quite independently, rediscovered it and realized its importance.

Mendelian or particulate inheritance, as now understood, incorporates five main concepts. It is not certain how far Mendel appreciated three of these, but two, segregation and independent assortment, he explicitly expressed in the form of two laws.

(i) Inheritance is particulate, and with the exception of the sex chromosomes (which determine sex, p. 48) the contribution from each parent to their offspring is equal. That is to say, the material determining the inherited constitution of the offspring, and transmitted through the sperm or egg, is not a miscible fluid. In fact it is composed of *genes*, which are discrete bodies.

They probably consist of giant molecules whose structure and spatial relationship with other similar molecules determines the nature of the inherited character.

(ii) The inherited factors do not contaminate one another. The genes are in pairs, one derived from each parent, and there may be thousands of such pairs in each organism. If the members of a pair are different, as they can be, they do not contaminate or alter each other's *structure* in any way, although they may alter each other's effects when together. Consequently the genes can be passed on to future generations in just the same form as they existed in the ancestors.

(iii) If two pure-breeding types which differ by a pair of contrasting characters are crossed and the resulting offspring or first filial generation (called F_1) are mated together, their progeny (F_2) will consist of some individuals genetically like the grandparents and others like the parents. In addition, among these progeny the forms will be represented in a particular ratio. Thus if we call one grandparental form A, the other B and the parental one C we will get F_2 individuals with the three possible forms in the proportions of 1A : 1B : 2C. In other words, the grandparental types reappear or *segregate* out in the F_2 generation, so that Mendel's first law is *The law of segregation*. This demonstrates the second proposition mentioned above.

(iv) Mendel also discovered that if two or more pairs of contrasting characters are present in the grandparents they will segregate in the F_2 (second filial generation) quite independently of one another. Genes, and therefore the characters, present together in one grandparent do not tend to stay together in the grandchildren. This rule is known as *The law of independent assortment*. Subsequent investigation has shown that it by no means always holds true (p. 41).

(v) The other important proposition is that genes are exceptionally stable. They do not readily change, or *mutate*, to a different form (called a *mutant*) although they can do so. Such a change may cause the appearance of a new inherited character. All genes do not have the same rate of spontaneous change. Even of those which mutate most often, only about 1 in 50,000 will do so in any one generation. Most do so even more rarely than this. In other words, of every 50,000 sex-cells produced, at the most only one will possess a newly arisen mutant form of any particular gene. However, as each such cell carries half the number of genes possessed by the parent (excepting some which are sex-linked, *see* p. 48), which must run into many thousands, a fair proportion of these cells will possess at least one new mutant of some sort. Of course these spontaneous changes can occur in any cell of the body and at any time, but unless they take place in a cell which will ultimately give rise to one or more sex-cells, the new mutant so formed cannot be inherited by the individual's offspring. Moreover, mutations tend to take place more often at certain stages in the formation of the sex-cells than at any other time or in any other type of cell. It will be seen from the accounts of Darwin's and Lamarck's ideas on inheritance (p. 23) that this rate of change by mutation is immensely smaller than anything they envisaged.

Before discussing Mendel's two laws, it is desirable to consider the material basis of heredity, which was not known when Mendel made his outstanding discoveries. The living cells which make up the tissues of an organism have within each of them a specialized part, the *nucleus*, which, amongst other things, apparently controls the growth and division of the cells. In the nucleus there are a number of thread-like bodies called *chromosomes*, which in normal circumstances can be seen only when the cell is dividing into two new cells. Even then they cannot be demonstrated except by using special microscopical techniques, or by killing the cell and staining it with certain dyes. If this is done, the chromosomes are seen as darkly staining bodies and their number (with certain exceptions) is the same in each cell of the body and also in all the members of any one species. The number of these chromosomes in a cell is known as the *diploid* number. For example, in some species of fruit-fly it is 8, and in man it is apparently 46, although until very recently it was believed to be 48. Careful investigations have shown that each chromosome has in the nucleus a partner like itself and each such pair of chromosomes differs from all the other pairs. This is true for all the chromosomes in the cell except that in one of the sexes, in bisexual organisms, there may be one pair whose members are not quite identical, and this pair is concerned with determining sex (p. 48).

Now in sexual reproduction the sex-cells, or *gametes*, as they are called, fuse at fertilization and form a single cell which develops into a new individual. If these gametes had the same number of chromosomes as other cells, the new individual formed at fertilization would have double this number, and, of course, in every subsequent generation the number would be doubled again. However, the method of production of the sex-cells ensures that this does not occur; during the cell divisions preceding their formation (called *meioses* or meiotic divisions), each chromosome of a pair comes to lie next to its partner. Eventually only one chromosome of each pair passes into each of the daughter cells, so that the gametes have half the number of chromosomes originally present (this number is the *haploid* number), and each original pair is represented by a single chromosome. No chromosomes are lost during this process, so that when one passes into one cell at division the other passes into the other daughter cell formed by the division. On fertilization the diploid number is restored and there are again two of each type of chromosome present.

The genes, which control inheritance, are carried in the chromosomes. Because the chromosomes are in pairs, each gene will be represented twice in a cell, once in one chromosome and once again at the same place, or *locus*, in the other member of the pair. Genes which are at the same locus are known as *allelomorphs* and can either be exactly alike or different.

Let us consider a single pair of contrasting characters in an organism such as the normal form of the Scarlet Tiger Moth, *Panaxia dominula* (L.), as compared with the rare variety *bimacula* (Fig. 1). The normal form carries two similar genes or allelomorphs controlling the distribution of black pigment, one in each of a pair of chromosomes. If we represent one allelomorph as *B*, the genetic constitution of *dominula* with regard to this character will be *BB*. We can represent its allelomorph producing the rare variety by *b*, and the genetic constitution or *genotype* of *bimacula* will be *bb* (the notation used here is simplified for the sake of clarity, p. 36). Such individuals in which the pair of allelomorphs are the same are called *homozygotes*. Now during the cell divisions which result in the formation of the

Fig. 1

Effects of crossing *bimacula* (a) and normal (b) forms of *Panaxia dominula*. Top row, parents. Middle row, the heterozygote, form *medionigra*, intermediate between the parents but variable. Compared with the normal form it has the central spot on the fore wings reduced or absent and an extra black spot on the hind wings. Bottom row, 1 : 2 : 1 ratio from a cross between two heterozygotes (*see text* p. 34).

gametes, only one of these allelomorphs will pass into each sex-cell, so that those from *dominula* will carry one *B* gene alone, whereas those from *bimacula* will carry one *b* gene also alone. At fertilization the diploid number of chromosomes will be restored by the fusion of egg and sperm and the new individual so formed will have the genetic constitution *Bb*. Such an individual which carries two different allelomorphs is referred to as a *heterozygote*. Moths of this genetic constitution are intermediate in appearance between *dominula* (Fig. 1b and j) and *bimacula* (Fig. 1a and g) and are called *medionigra* (Fig. 1 c, d, e, f, h, and i).

Because all the gametes produced by a *BB* individual will carry one gene *B*, and all those from a *bb* individual will carry one *b*, all the offspring produced by crossing them must be *Bb*. However, an individual of the genetic constitution *Bb* will produce in equality gametes carrying either *B* or *b*, but not both. Which any particular gamete is carrying will depend on which of the pair of chromosomes it possesses. As only one of each pair of chromosomes passes into each of the two daughter cells at cell division, one gamete will have *B*, the other *b*. Consequently, equal numbers of those carrying *B* and those carrying *b* will be produced. If two heterozygotes are mated together the male will produce equal numbers of sperm carrying *B* and sperm carrying *b*, and the female will produce equal numbers of eggs of these two types. Consequently, a *B*-carrying sperm will have an equal chance of fertilizing a *B*- or *b*-carrying egg, so that a large number of such fertilizations will produce, on an average, equal numbers of *BB* and *Bb* individuals. Using the same argument, *b*-carrying sperms will produce equal numbers of *Bb* and *bb* offspring; by combining the results from the two types of sperm we find that offspring will be produced in the proportions 1*BB* : 2*Bb* : 1*bb*. It will be noted that the two homozygotes are genetically like the grandparents, whereas the heterozygotes are like the parents, thus the grandparental types have segregated out in the F_2 generation.

The results of any such mating between two heterozygotes can be determined by constructing a table in the form of a square. Along one side one can place the types of sperm produced, along the other the types of egg, and by following along the columns and lines one can determine the genetic constitution of the offspring of the union of any two types by looking in the appropriate compartment. By constructing other such squares,

		Bb male sperm	
		B	b
Bb female	B	BB	Bb
eggs	b	Bb	bb

offspring

it will at once become obvious that two similar homozygotes, let us say BB, can only produce offspring like themselves. Two different homozygotes will produce offspring all of which are heterozygotes, in this instance Bb, and heterozygotes mated together will produce $1BB : 2Bb : 1bb$ (*see* diagram above). The other possible mating is a cross between a heterozygote and a homozygote which will give equal numbers of heterozygotes and homozygotes, as can be seen from the square diagram below. A mating of this sort is known as a *backcross* because the F_1 is crossed back to one of the parental genotypes.

		Gametes from heterozygote Bb	
		B	b
Gametes from homozygote BB	B	BB	Bb
	B	BB	Bb
	Total offspring	$2BB$	$2Bb$

Dominance

In the example just cited the heterozygote is intermediate in appearance between the two homozygotes. This situation, however, is the exception rather than the rule. In most crosses between individuals showing sharply contrasting characters controlled at a single locus, the heterozygote is indistinguishable from one of the homozygotes. In other words, one of the allelomorphs exercises its full effect regardless of whether it is partnered by an identical or different one. A character which has this attribute is said to be *dominant*, whereas that which only appears if the individual is homozygous for the allelomorph controlling it is called the *recessive*. When Mendel worked with peas, he investigated a pair of contrasting characters, green *versus* yellow seeds. In this instance yellow is dominant to green. By convention, when assigning a letter to represent a pair of allelomorphs, we give a capital to that which produces the dominant, and a small letter to the allelomorph determining the presence of the recessive variety. Consequently, for green *versus* yellow, we can denote the allelomorph producing green by g and that producing yellow by G. A single locus is always denoted by the same letter or letters. Consequently a particular set of allelomorphs, since they are all to be found at a particular place or locus on a certain chromosome pair, are always denoted by the same letter or letters with appropriate changes in case (i.e. G or g), or with the addition of suitable letters or numbers to distinguish different allelomorphs. In the previous example we should have used B^D and B^B instead of B and b because one character was not dominant to the other, but did not do so because this more complicated notation might have confused the beginner; in the example of Mendel's peas, where there is dominance, we use G and g.

Mendel used two pure breeding lines of peas, one with green seeds gg, the other with yellow ones GG. One line could only produce gametes carrying g and the other gametes carrying G, because both were homozygous. Consequently the first filial generation all had the genotype Gg, but because there was complete dominance all had yellow seeds like their GG parent. The F_2 generation was composed of individuals in the proportions $1GG : 2Gg : 1gg$, as would be expected from our previous example. However, because there was dominance, the heterozygotes had yellow seeds like their parents and one of their grandparents. Thus there were only two types of seeds judged by their appearance, and these were in the ratio of 3 yellows : 1 green. The characters displayed by an organism are called its *phenotype*, whereas its genetic constitution is called the *genotype*. Thus in Mendel's crosses there were only two phenotypes in the ratio of 3 yellow : 1 green, but three genotypes in the proportions $1GG : 2Gg : 1gg$. One of the classes with yellow seeds was composed of homozygous *GG* individuals which would breed true if mated to individuals like themselves, and the other of heterozygous *Gg* individuals whose offspring would segregate.

By crossing two pure-breeding strains which differ by a pair of contrasting characters, we can determine whether one character is dominant or not by the proportions of the types in the F_2 generation. If there is dominance, we will get two phenotypes in the proportions 3 : 1, the dominant class being the commoner one. On the other hand, if there is no dominance, we will get a 1 : 2 : 1 ratio, with the heterozygotes the commonest class.

When there is dominance one will get from a backcross a 1 : 1 ratio of heterozygotes to homozygotes as explained on page 35. However, if the heterozygote is mated to the dominant homozygote, all the offspring will have the same appearance although only half of them will be homozygotes. On the other hand, if it is paired with the recessive form the offspring will consist of about equal numbers of the dominant and recessive forms. To take an actual example, a pink-shelled individual of the snail *Cepaea nemoralis* (L.) (*Yy*) mated to a yellow-shelled recessive form (*yy*) produced 45 pinks and 33 yellows.[8]

It will be noted that the expected 1 : 1 ratio is not exactly obtained although the numbers do approach the correct value. In fact, in breeding experiments one will not often get the expected proportions just as one will not often get exactly the same number of heads and tails on tossing a coin for a limited number of throws, even though the expected number of heads is exactly half the number of throws. One can think of the outcome of fertilization in much the same way as one can think of the outcome of throwing coins. If there are two kinds of sperm, *A* and *a*, and one kind of egg, *a*, as in a backcross, then, just as with throwing a coin and expecting half heads and half tails, each egg will have an equal chance of being fertilized by *A* or *a* sperm. As explained before (p. 34), one will therefore expect in such a cross equal numbers of the two genotypes *Aa* and *aa*. However, exact equality will seldom be obtained for the same reason that equal numbers of heads and tails will only rarely be obtained in a tossing experiment (*see* p. 116). For by chance alone one will usually get either more A or more a sperm fertilizing the eggs, just as by chance one will either get more heads or more tails as a result of throwing coins.

Multiple allelomorphs

We have been considering pairs of allelomorphs. It is, however, possible for genes to exist in more than two forms, in which case the set of them constitutes a multiple allelomorphic series; thus in the snail *Cepaea nemoralis* (Fig. 5) there are several distinct shell colours, including brown, pink and yellow.[8] The allelomorphs controlling these three behave as if they are at a single locus. Because the chromosomes are in pairs, no more than two allelomorphs can be present in an individual nucleus. Consequently, if there are three possible forms of the gene at one locus, for example Y^B, Y, y, as in *Cepaea*, then the only possible genotypes are the homozygotes Y^BY^B, YY, yy, and the heterozygotes Y^BY, Y^By, Yy.

Independent assortment

Mendel did not work only with one pair of allelomorphs at a time, but, on occasion, with two or more sets of contrasting characters simultaneously. For example amongst others he investigated green *versus* yellow seeds in peas, and wrinkled *versus* smooth ones, in which smooth is dominant. He crossed plants having yellow smooth to those with green wrinkled seeds. In such a cross we can denote the allelomorph producing wrinkled (the recessive) by w and that controlling smooth by W. The genotype of the yellow smooth line was $GGWW$ and of the other was $ggww$. Now the sex-cells produced by $GGWW$ will include only one of each pair of chromosomes, and therefore will carry only one G and one W gene. We can therefore denote the gamete by GW, and that from the other line, using a similar argument, must be gw. On fertilization the genotype will be $GgWw$: that is to say, it will be heterozygous at two loci, and because of the dominance relationships the phenotype will be yellow smooth. These individuals, being heterozygotes, will produce, in equal numbers, sex-cells carrying G or g, and simultaneously W or w. As each must carry one gene from each pair, there are four possible combinations, GW, Gw, gW, gw. If there is no tendency for the grandparental allelomorphs G and W or g and w to be held together—and there will not be if they are on different chromosomes—then the four kinds of gametes will be produced in equal numbers. As before, any sort of pollen is equally liable to meet any sort of egg. We can therefore determine the genetic constitution of the various offspring from a cross between the double heterozygotes by constructing a square diagram as before :

Gametes

	GW	Gw	gW	gw
GW	$GGWW$	$GGWw$	$GgWW$	$GgWw$
Gw	$GGWw$	$GGww$	$GgWw$	$Ggww$
gW	$GgWW$	$GgWw$	$ggWW$	$ggWw$
gw	$GgWw$	$Ggww$	$ggWw$	$ggww$

Gametes

Extracting the different genotypes from the body of the table we find:

Phenotypes

Yellow smooth		Yellow wrinkled		Green smooth		Green wrinkled	
$GGWW$	1	$GGww$	1	$ggWW$	1	$ggww$	1
$GGWw$	2	$Ggww$	2	$ggWw$	2		
$GgWW$	2						
$GgWw$	4						
	9		3		3		1

If the genotypes are examined it will be found that there are $4WW$'s : $8Ww$'s : $4ww$'s; that is, they are in the expected ratio of $1 : 2 : 1$, and the same is true for the three genotypes GG, Gg and gg. However, because there is dominance in this example, the ratios of phenotypes will be a 3 : 1, and not a 1 : 2 : 1 for each character. The observed ratio of the four classes of phenotypes, taking into account both characters, is 9 : 3 : 3 : 1. This is characteristic of the F_2 generation of a cross between different pure-breeding lines when two pairs of contrasting characters are assorting independently of one another and there is dominance. This is Mendel's principle of independent assortment, but as will be seen in the next chapter it does not always apply.

III

SOME OTHER MENDELIAN PRINCIPLES

Linkage

IN MENDEL'S breeding experiments, the factors investigated assorted independently, but it will be remembered that the genes are carried in the chromosomes and therefore their distribution in the gametes is controlled by the distribution of the chromosomes. Consequently if two loci are close together in the same chromosome, they will be carried together into the same sex-cell and will therefore not assort independently. Although Mendel did not discover this phenomenon, it was encountered in breeding experiments soon after the rediscovery of his paper.

To illustrate the principle, we can consider two pairs of contrasting characters in the snail *Cepaea nemoralis* (Fig. 5). There are a number of colour varieties of the shell, and two of them, pink and yellow, are controlled by a pair of allelomorphs with pink dominant, as has been mentioned earlier. We can call these Y and y. The shell can also either be banded (Fig. 5, a, c, d, e and f) or unbanded (Fig. 5b), the unbanded condition being dominant.[8] The genes controlling this pair of characters we will call U and u. If we cross a pink unbanded snail of the constitution $YYUU$ with a yellow banded one $yyuu$, we can get only the double heterozygote $YyUu$ among the offspring. Now, because the loci involved are very close together on one chromosome, Y and U will be in one member of the chromosome-pair, and y and u in the other, the former derived from one parent, the latter from the other. If we represent the chromosome involved by a line, then we can mark the genes in it with letters, so that the two parents, each with a pair of chromosomes, will be represented as $\frac{Y\ U}{Y\ U}$ and $\frac{y\ u}{y\ u}$. When these produce gametes, one chromosome will pass into each cell so that the gametes from the two

41

forms can be represented as $\underline{Y\ U}$ and $\underline{y\ \ u}$ respectively. On fertilization the chromosome number is restored by the fusion of these sex-cells and we get the genotype $\dfrac{Y\ U}{y\ \ u}$ which has a pink unbanded shell. An individual of this constitution, when it is forming sex-cells, will again pass one chromosome into one, and the other chromosome into the other cell at each division, so that the gametes will be either $\underline{Y\ U}$ or $\underline{y\ \ u}$. If such an individual is mated with another like itself, we can ascertain the types of offspring produced with a square diagram as before. They will be $1\ \dfrac{Y\ U}{Y\ U} : 2\ \dfrac{Y\ U}{y\ \ u} : 1\ \dfrac{y\ \ u}{y\ \ u}$. In other words, we get a 1 : 2 : 1 ratio of genotypes, and, because of the dominance, we have 3 pink unbanded snails : 1 yellow banded. If the genes had assorted independently instead of being linked together on the same chromosome, we would have obtained 9 pink unbanded : 3 pink banded : 3 yellow unbanded : 1 yellow banded.

If the original grandparents had been pink banded and yellow unbanded, the chromosomes in the F_1 generation would have been $\underline{Y\ \ u}$ and $\underline{y\ \ U}$. Consequently, in the F_2 generation we would have got 1 yellow unbanded : 2 pink unbanded : 1 pink banded. Therefore the results in the F_2 differ, depending on the combination of genes in the grandparents. If two genes, controlling dominant characters, are in the same chromosome they are said to be in *coupling*, and a 3 : 1 ratio is obtained. If a dominant and a recessive are together, they are said to be in *repulsion*, and a 1 : 2 : 1 ratio appears.

Crossing-over

From the previous example it might seem that genes in the same chromosome must always be inherited together. This, however, is not strictly true, because during the cell divisions which result in the halving of the chromosome number and the production of sex-cells, genetic material may be exchanged between the two chromosomes of each pair.

In order to understand how this is brought about, it is necessary to consider the process of gamete formation in more detail than was given on p. 32. In the early stages of meiosis (p. 32) each member of a pair of chromosomes comes to lie next to, and twist round, its partner in such a way that each gene is intimately associated with its allelomorph on the other chromosome. The chromosomes then divide longitudinally so that each consists of two identical strands. Consequently, all the genes in a particular chromosome are now represented twice, one set being found in each strand. In an individual with a pair of chromosomes of the constitution $A\ B$ and $a\ b$, the relationships of the strands of the chromosome pair will come to be

$$\frac{\overline{\begin{array}{cc}A & B\end{array}}}{\overline{\begin{array}{cc}A & B\end{array}}} \quad \overline{\begin{array}{cc}a & b\end{array}} \quad \overline{\begin{array}{cc}a & b\end{array}}$$. At about this time one or more of these four threads tends

to break at a point, or points, along its length. This happens in such a way that if a break occurs at a particular place in one thread, a similar break will appear at the same place in one of the other three strands which make up the pair of chromosomes. The ends rejoin in such a way that part of one thread becomes interchanged with a similar part of the other. For example if a break occurs between genes A and B, in one, and in an exactly similar place between a and b, in another, the chromosomes will

have the following relationship: $\frac{\overline{\begin{array}{cc}A & B\end{array}}}{\overline{\begin{array}{cc}A & B\end{array}}} \quad \frac{\overline{\begin{array}{cc}a & b\end{array}}}{\overline{\begin{array}{cc}a & b\end{array}}}$. When the broken ends rejoin,

the piece carrying b will combine with that carrying A, and the piece carrying a with that carrying B. Consequently two new chromosome threads, $A\ \ b$ and $a\ \ B$, are formed. They will

have the following relationship to one another: $\begin{array}{c}A\ \ B \\ A \diagdown B \\ a \diagup b \\ a\ \ b\end{array}$. The

position where the two threads cross over is called a *chiasma*.

In our example, each chromosome of the pair will now be

composed of two strands; one chromosome will have A B and A b, thus $\dfrac{A\ B}{A\ b}$, and the other will have a b and a B, thus $\dfrac{a\ b}{a\ B}$. The chromosomes then move apart and pass to opposite ends of the cell, which divides to form two new cells with one of the pair of chromosomes in each. The two daughter cells then divide again without any further division of the chromosomes. During this division the two threads part from one another to become chromosomes in their own right, and one passes into each new cell. Consequently we now have four cells of the constitution A B, A b, a B and a b, the first two derived from one and the last two from the other product of the first division. In this description of *meiosis* (the two cell divisions directly preceding gamete formation) we have been considering the fate of the products of only one pair of chromosomes but the same sequence of events, of course, applies to all the other pairs the cell possesses.

Had the organism been heterozygous for A and a, but homozygous for B, we would still get four sex-cells, but only two types, A B and a B, since the exchange of B with B leaves the situation unaltered. In other words, when we are considering only one locus on each pair of chromosomes, as in the cases of segregation and independent assortment, discussed in the previous chapter, we can disregard this complicated process and merely say that a heterozygote Aa will produce equal numbers of A and a gametes.

The fate of the four cells at the end of meiosis differs according to the circumstances. In male animals each forms a sperm, whereas in females three of them degenerate and only one survives to form an egg-cell. In the higher plants, on the other hand, meiosis gives rise to cells which form either *megaspores* or *microspores*, which produce respectively embryo sacs with ovules and pollen grains. However, what concerns us here is that meiosis can result in the formation of chromosomes which differ from those of the individual's parents by allowing some genetic material from one to be exchanged with its partner chromosome.

Now the chance of a break occurring between any two loci will be increased the further apart they are on the chromosome. If they are extremely close together, there is very little chance of a break occurring between them, as in the example taken from *Cepaea*. If they are at opposite ends of the chromosome, at least one, and probably more than one, break will appear between them at meiotic cell division.

The process by which allelomorphs are exchanged between *homologous chromosomes* (chromosome pairs) is referred to as *crossing-over*, and the classes of offspring resulting from the new combinations are called cross-over classes. If we take two individuals, *AABB* and *aabb*, with the loci not linked, the double heterozygote will be *AaBb*. If we now cross this to the double recessive *aabb*, we will get 1*AaBb* : 1*Aabb* : 1*aaBb* : 1*aabb*. If, however, the two genes are very closely linked in coupling and there is no crossing-over the only classes produced will be 1*AaBb* : 1*aabb* because the gametes produced by the double heterozygote will carry only $\overline{A\ B}$ or $\overline{a\ \ b}$ (*see* above). If, however, there is some crossing-over, a break will occur between *A* and *B* in some cells but not in others, so that in a small proportion of the gametes $\overline{A\ \ b}$ and $\overline{a\ \ B}$ will be found, and these types of sex-cells will be in equal numbers since one of each results from a single cross-over. Consequently amongst the offspring there will be all four possible genotypes, but the majority will consist of *AaBb* and *aabb* individuals in equal numbers, and the minority of *Aabb* and *aaBb* again in equal numbers. For example we might get the following:

Proportion of gametes from

double heterozygote $\dfrac{A\ \ B}{a\ \ b}$

Gametes from	5*AB* :	1*Ab* :	1*aB* :	5*ab*	
recessive $\dfrac{a\ \ b}{a\ \ b}$ all *ab*	5*AaBb*	1*Aabb*	1*aaBb*	5*aabb*	offspring

The *cross-over value* is calculated by taking the percentage of all individuals produced which are cross-over classes. In this example the cross-over value will be:

$$\frac{1Aabb + 1aaBb}{5AaBb + 1Aabb + 1aaBb + 5aabb} = \frac{2}{12} = 16.67\%.$$

It is obvious that the further apart the genes in the chromosome, the greater must be the cross-over value, so that by obtaining various cross-over values the order of the genes in the chromosome can be determined, together with some measure of their relative distances apart. A cross-over value of 50% is equivalent to independent assortment and cannot normally be exceeded. Therefore if the genes are sufficiently far apart on the chromosome they cannot be distinguished from genes on different chromosomes unless a gene lying between them is shown to be linked to both.

The principles involved can be illustrated by two actual examples taken from the fruit-fly *Drosophila melanogaster*. A fly heterozygous for the genes for normal colour *E versus* a black body *e* (ebony) and short thick bristles *Sb* (stubble) *versus* normal ones *sb* (i.e. *Sbsb Ee*) was crossed with the normal-bristled black-bodied double recessive (i.e. *sbsb ee*). The genes were in coupling and the offspring were 85 *Sbsb Ee* (stubble, normal colour) : 14 *Sbsb ee* (stubble, black) : 13 *sbsb Ee* (normal bristles, normal colour) : 88 *sbsb ee* (normal bristles, black colour). The cross-over classes are stubble black and non-stubble normal colour, so that the cross-over value is 13·5%. A similar cross was made between an individual heterozygous for stubble and for veinlet (*ve*), a gene interrupting the veins of the wings. The double heterozygote *Sbsb Veve* was crossed to the double recessive *sbsb veve*, and the resulting progeny were 42 *Sbsb Veve* (stubble, normal) : 29 *Sbsb veve* (stubble, veinlet) : 29 *sbsb Veve* (normal bristles, normal wings) : 48 *sbsb veve* (normal bristles, veinlet). The cross-over classes are *Sbsb veve* and *sbsb Veve*, so that the cross-over value is 39·2%. Because this last value is larger than that between stubble and ebony, the locus controlling veinlet (*ve*) must be further away from stubble (*Sb*) than that controlling ebony (*e*). There are therefore two possible orders of the loci *Sb   e   ve* or alternatively *e   Sb   ve*. That is to say *Sb* can either be between *ve* and *e* or not between them.

A cross-over value between ebony and veinlet would fix the order of the genes on the chromosome. If this cross-over value were greater than that between *Sb* and *ve* (i.e. in the region of 50%) ebony would have to be the far side of stubble, the order being $\dfrac{e \qquad Sb \qquad ve}{\leftarrow 13\cdot5 \rightarrow \leftarrow -39\cdot2 \rightarrow}$ whereas if the cross-over value between *ve* and *e* were less than between *Sb* and *ve*, *e* must be nearer to *ve* than is *Sb*, and, therefore, lie between the two, the order being

$$\dfrac{\overset{\longleftarrow\text{------}39\cdot2\text{------}\longrightarrow}{ve \qquad e \qquad\quad Sb}}{\leftarrow-13\cdot5\rightarrow}$$ (this is, of course, the same as Sb   e   ve).

In fact the cross-over value between the *ve* and *e* loci is so close to 50% that no linkage could be detected with certainty. Thus the order of the genes on the chromosome must be e   Sb   ve.

It will be seen that the sum of the cross-over values between *e* and *Sb* and between *Sb* and *ve* exceeds 50% but that the value between *e* and *ve* is not greater than 50%. This follows from the fact that more than one cross-over may occur in so long a section of chromosome. If *e* and *ve* are on the same chromosome (in coupling), and there are an even number of cross-overs between them, they will be inherited together since a second cross-over will restore the combination disrupted by the first. Only if there is an odd number of cross-overs between the two will cross-over classes appear among the offspring. Consequently the cross-over value between two loci is nearly always less than the sum of the values between them and some other locus lying between them.

By using such breeding methods it is possible to find the linear order of all the known genes in the chromosomes, and also to determine which groups of genes are unlinked to any others, and, therefore, lie in a different pair of chromosomes. The results obtained from such procedures are known as chromosome maps because the relative positions of the genes are mapped. They have been constructed for a few animals and plants with varying degrees of completeness. At the moment the chromosome map of man is beginning to be constructed, but the process is inevitably slow when experimental breeding cannot be practised.

Sex-linkage

As was mentioned earlier, each chromosome has a partner like itself and the pair differs from all the other chromosomes in the cell. In organisms, however, where the sexes are separate and sex is determined by genetic factors, as in some plants and most animals including man, one, or occasionally more than one, pair is composed of chromosomes that are not identical. Such chromosomes determining sex are called *sex chromosomes*, as distinct from *autosomes*, in which the pairs are similar. When sex chromosomes are present, one sex will have two similar chromosomes, which, by convention, are referred to as X chromosomes, whereas the other will have two non-identical ones, one of which will be an X and the other a Y. The difference between them springs from the fact that one or both of these types have genetic material present in them which is not represented in the other, so that they are often of different length and do not pair completely during meiosis. A gene in these non-pairing regions is said to be sex-linked.

In mammals, including man, the female has the constitution XX and is called the *homogametic sex*, and the male is XY and is the *heterogametic sex*. In birds, on the other hand, the female is XY and the male XX. In mammals the female produces only gametes carrying an X but the male produces two types in equal numbers, those carrying X and those carrying Y. When fertilization is due to an X-carrying sperm, the offspring will be female, when due to one with a Y it will be male. Thus in man it is the genetic constitution of the sperm which determines the sex of the embryo, whereas in birds it is that of the egg which does so.

		XX female eggs		
		X	X	
XY male sperm	X	XX	XX	females
	Y	XY	XY	males

Because there is only one type of gamete from the homogametic sex, but two, formed in equal numbers, from the heterogametic sex, about equal numbers of the two sexes will usually be obtained at fertilization, although in later life the sex ratio may not remain the same because of differential mortality between the two sexes.

Now any gene in the non-pairing part of the Y chromosome (the part not represented in the X) will never be able to cross over into an X, and will, therefore, pass in mammals from father to son, and in birds from mother to daughter. However, a gene on that part of the X which does not pair with the Y can be found in either sex, for the X chromosome is found in both. The allelomorph determining the presence of red-green colour-blindness in man is at such a locus, and the character is recessive. Consequently a colour-blind woman will have the constitution X^bX^b, where X represents the chromosome and b the allelomorph in it. If she marries a man with normal colour vision, he will have the constitution X^BY, for there is no B locus in the Y chromosome. The gametes produced by the female will all be of the constitution X^b and those of the male either X^B or Y. Consequently the offspring of such a marriage will be X^BX^b females, which will have normal vision, as colour-blindness is recessive, and X^bY males who will be colour-blind as there is no B gene to suppress the effects of the b allelomorph. Thus all the males will be colour-blind and all the females will be heterozygous for the character, but will have normal vision. If one of these heterozygous daughters marries a normal man, on the average half of her sons will be colour-blind and half normal. Half of her daughters will be heterozygotes and half will not be carrying the gene for colour-blindness, as will be seen from the diagram.

		X^BX^b female eggs	
		X^B	X^b
X^BY male sperm	X^B	X^BX^B	X^BX^b	females
	Y	X^BY	X^bY	males

If the colour-blind son marries a normal woman all his sons will be normal and will not carry the gene for colour-blindness, whereas all his daughters will be heterozygous for it, as will be seen from the next diagram. In other words, a sex-linked recessive controlled by a gene present on the X chromosome in the

X^BX^B normal female
eggs

		X^B	X^B	
Colour-blind	X^b	X^BX^b	X^BX^b	females
male X^bY sperm	Y	X^BY	X^BY	males

heterogametic parent will only be inherited by the homogametic offspring.

Darwin was conversant with many of the effects of sex-linkage for colour-blindness and haemophilia (a blood disease) in man. He wrote 'generally with the haemorrhagic diathesis, and often with colour-blindness, and in some other cases, the sons never inherit the peculiarity directly from their fathers, but the daughters, and the daughters alone, transmit the latent tendency, so that the sons of the daughters alone exhibit it. Thus, the father, grandson, and great-great-grandson will exhibit a peculiarity,—the grandmother, daughter, and great-granddaughter having transmitted it in a latent stage.'

Sex-controlled inheritance

Sex-controlled (sex-limited) inheritance must not be confused with sex-linkage. In many animals characters are known which appear in one sex only. This does not spring from the fact that they are controlled by a gene on a sex chromosome, for the gene responsible may be, but often is not, sex-linked. The sex-control results from the fact that only one of the two sexes provides the condition in which the gene can exert its effect. The phenomenon is particularly evident in mimetic butterflies, which are discussed in Chapter X. In man premature baldness is a sex-controlled dominant, the character normally only appearing in males despite the fact that the locus responsible is not on a sex chromosome.

Cytoplasmic inheritance

The *cytoplasm* is the part of the living cell which is outside the nucleus. The nucleus is often approximately spherical in shape and it contains the chromosomes. Now an additional method of inheritance which is sometimes found in plants and also in animals is due to factors carried in the cytoplasm and not in the nucleus, and is therefore called cytoplasmic inheritance. One can usually test if variation is inherited in this way by making reciprocal crosses. These are crosses in which the sex of the individuals of two types is reversed in one of a pair of matings. For example if the female was *AABBCC* and the male *aabbcc*, the reciprocal cross would be female *aabbcc* with male *AABBCC*. In these circumstances, the resulting progeny would be the same if only Mendelian factors were involved, providing they were not sex-linked, whereas the result would be quite different if there were cytoplasmic inheritance. The offspring would usually be much more like the female parent. This results from the fact that the two sexes contribute different amounts of cytoplasm to the offspring, the female parent usually contributing much more (a million times more is not unusual), so that the offspring would tend to be more like it than like the male.

Mutation

In the foregoing discussion we have seen that different allelomorphs can be found at a particular locus, but only brief mention has been made as to how these arise. The change of a gene from one form into another is known as a mutation, and the result of such a change is called a *mutant*. Reverse mutation is also known to occur: that is to say, a mutant can change back to its original form.

Little is known of the structure of the gene, or of how it exerts its effect in development. It is clear, however, that although gene-mutation is sometimes due to the loss of genetic material, it is probably more often due to a chemical change in the material of the gene. The phenomenon of reverse mutation indicates that the change cannot always result from a loss, for if this were so, material would not be available allowing the mutant to change back to its original form. Spontaneous changes occur very rarely at most loci, and a new mutant seldom appears at any one locus more often than once in 25,000 offspring: that is to say, once in 50,000 chromosomes per generation, for each individual has, in every cell, two chromosomes carrying the locus. Mutation is usually far less common than this.

Chemical substances, such as colchicine or nitrogen mustard, are known which increase the rate of production of such mutants (the mutation-rate). Ionizing radiation can also increase the rate. It seems clear, however, from available evidence, that natural radiation does not account for all the observed mutants that occur in the wild. This would be expected, for it cannot be supposed that the large molecules of which the genes are composed are completely stable in the chemical sense, and will never change except under the influence of ionizing radiation.

The unexpectedly low frequency of mutation, considering the size of the molecules involved, is probably the result of natural selection, and it is known that genes can influence the mutation-rate of other genes in the organism. This low mutation-rate is in remarkable accord with the requirements of evolution in organisms, in which inheritance is particulate, but is impossibly low for any evolutionary hypothesis postulating blending inheritance. In this connection it is a significant fact that mutation-rates per generation are very constant through a large range of organisms. If the rate is measured in chronometrical time, however, mutation-rates vary wildly, but it is the generation and not the length of a generation that is important from this point of view in evolution, thus suggesting that mutation-rates in any organism have been evolved in accordance with its requirements.

Chromosome mutations

Besides gene mutations, or point mutations as they are called, other changes can occur which result in a new character appearing and being inherited. These include many types of chromosome change,[73] among which are (i) *deletion*, that is, the loss of part of a chromosome, which, if the piece lost is large, is usually lethal to the individual which inherits it ; (ii) *duplication*, which is the presence of a similar small piece of a chromosome in two places in one chromosome. Consequently one or more loci will be represented twice, thus *ABCDCDEFG* ; (iii) the exchange of parts of a chromosome from a member of one pair to that of another, known as a *translocation*; (iv) the breakage of a chromosome in two places, the rotation of the broken piece through 180°, and the rejoining of the broken ends, called an *inversion*. When this happens the order of the genes along the chromosome is reversed over the length of the broken piece. For example, if the normal order is *ABCDEFG* and breaks occur between *B* and *C*, and *E* and *F*, and the part with *CDE* is rotated through 180° and joins up again, the order of the genes on the new chromosome will be *ABEDCFG*. During meiosis in the heterozygote for an inversion, the chromosomes have to loop about one another when an inversion is present, in order to pair properly. Such loops are quite characteristic and can be recognized in certain chromosomes (Fig. 2). Moreover, crossing-over in the region is partially suppressed for mechanical reasons and also the products of the few cross-overs tend to be destroyed. Consequently inversions were first recognized as supposed 'genes' suppressing crossing-over.

Mutations can occur in the reproductive organs or in other parts of the body at any stage in development. If they do not occur in the cells that ultimately give rise to gametes, the mutant gene cannot be inherited by future generations but may cause the body of the individual carrying it to show a mosaic pattern of characters, the cells not carrying the mutant being normal and those carrying it showing the mutant characters.

Polyploidy

Not only can the structure of individual chromosomes change, but the actual number of chromosomes present may alter through some mischance at cell division. Thus a chromosome can be lost or an extra one gained. The loss of a chromosome

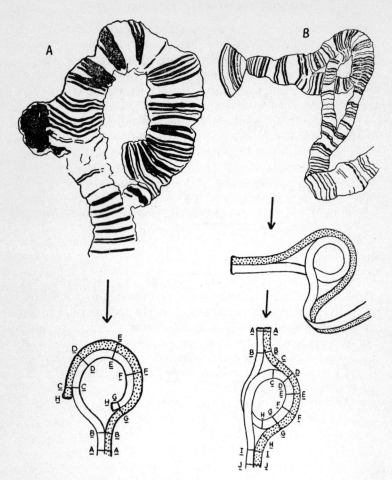

FIG. 2

Parts of two different pairs of the giant salivary-gland chromosomes of a single larva of the midge *Chironomus tentans* F. heterozygous for two different inversions. A, loop caused by an inversion near the end of one of a pair of chromosomes. Except where the inversion disturbs the situation, the two chromosomes are so closely associated that they appear as one. The spatial arrangement of the chromosomes is explained in the diagram, in which corresponding arbitrarily chosen loci on the two chromosomes have the same letter. The inversion involves the section *CDEFG*. B, a similar loop not close to the end of its chromosome, involving a region *CDEFGH* as shown in the accompanying diagram.

is usually far more serious for the viability of the organism than the acquisition of an extra one. Perhaps one of the most important numerical changes of this type in evolution is that in which the whole chromosome complement is doubled. An animal or plant in which more than one pair of each type is present is called a *polyploid* to distinguish it from a *diploid*, the normal state. If each chromosome is represented three times, the individual is said to be a *triploid*, if four times a *tetraploid*, and so on. The usual way in which such polyploidy is formed is by the failure of cell division during gamete formation so that the sex-cell still has the diploid and not the haploid chromosome number. If fertilization took place between such a gamete and a normal one, a triploid would be formed, whereas if two abnormal gametes of this type fused a tetraploid would result.

Polyploids, like other types of mutant, can produce definite effects in an organism. Many polyploids have larger and sometimes fewer cells than their diploid parents. Moreover, sometimes, but not always, the individual is bigger than the diploid. Another character common to most polyploids is that they tend to be very much less fertile, and with triploids this approaches absolute sterility. In animals this results from the fact that in certain forms of polyploid the genetic balance of the organism is upset so that the normal distinction between male and female is not maintained and sterile intersexes occur. When this does not happen, there is still abnormal pairing of the chromosomes during meiosis, so that gametes having various numbers of chromosomes are formed, and these usually fail to produce viable offspring and may themselves not develop properly. Consequently polyploidy is rare in animals and is chiefly confined to species where bisexual reproduction has been dispensed with, for if there were only bisexual reproduction the polyploids would be at a great disadvantage. Polyploids are chiefly found in animals among the parthenogenetic turbellaria (flatworms), crustacea, mollusca, insecta and oligochaeta (earthworms).[73]

In plants polyploidy has been very important in evolution. However, here again there is a reduction of fertility, both because of physiological upsets in reproduction and because of abnormal pairing of the chromosomes during meiosis. There exist two main types of polyploids, *autopolyploids* and *allopolyploids*. An autopolyploid is formed by a doubling of the chromosome number of any individual of any species of organism. In them fertility tends to be much reduced because there are then more than two of each type of chromosome in each cell. For example if a diploid species had four pairs of chromosomes which we can call AA, BB, CC, and DD its autotetraploid would have AAAA, BBBB, CCCC and DDDD. Consequently in tetraploids, groups of three and one, or four, chromosomes, as well as those of two and two, may be formed by each of the types of chromosome at meiosis, so that the gametes have very irregular numbers of chromosomes and do not give rise to offspring. If there is a second species with, for example, four chromosome pairs RR, SS, TT and UU and a hybrid between this and the first species is formed, it will have the chromosomes A, B, C, D, R, S, T and U. If the chromosomes A, B, C and D are very different from R, S, T and U there will be little or no pairing at meiosis and the hybrid will be very infertile. If the chromosome complement of such an individual is doubled an allopolyploid (allotetraploid) will be formed and will have the chromosome pairs AA, BB, CC, DD, RR, SS, TT and UU. As a result of this, each type of chromosome now has one and only one partner like itself, so that pairing at meiosis is normal and fertility is likely to be much improved. In fact when a hybrid is sterile because the chromosomes of one of the parent species are different from those of the other, and therefore there is no pairing in the hybrid at meiosis, fertility can be restored if the chromosome number is doubled. Thus allopolyploids tend to be more fertile than autopolyploids. Naturally two species do not have to have the same haploid number of chromosomes for them to be able to form an allopolyploid.

The effect of polyploidy on fertility is important from the evolutionary point of view since it very much reduces or even eliminates the possibility of the polyploids exchanging genes with their diploid ancestors, because the resulting progeny will mostly be triploids, which are more or less sterile. This aspect of polyploidy will be discussed in more detail in Chapters XI and XII. It will suffice to mention two further points here. Because polyploids, even allopolyploids, have reduced fertility, they will seldom survive in organisms that reproduce only by sexual means.

Thus polyploids are commoner in plants than in animals, particularly those plants which can perpetuate themselves by tubers, runners, or some other asexual means. As previously pointed out, polyploidy in animals is chiefly confined to asexually reproducing forms, or those that are *parthenogenetic*. In such animals, the egg develops without the sperm nucleus being fused with that of the egg, so that males are usually dispensed with.

Mutants in natural populations

We have discussed how genes segregate in the progeny of a cross, and it is now necessary to apply this knowledge to populations. It was G. H. Hardy who in 1908 laid the foundation of modern population genetics. He investigated mathematically the distribution of the genotypes in large populations in which there was no selection operating, and in which there was a pair of allelomorphs A and a in the ratio $p : q$, where $p + q = 1$. If the allelomorphs are not sex-linked and there is random mating, that is, the choice of a mate is not determined by its genotype with respect to A and a, the frequency of the genotypes AA, Aa and aa in the next generation will be $p^2 : 2pq : q^2$. Hardy showed that in subsequent generations this ratio will not change.

This rule forms the foundation of all population genetics, and because it will be mentioned again and again in subsequent chapters it is worth showing here that it is true. If we represent the frequency of the homozygote AA by X, the heterozygote Aa by Y, and the homozygote aa by Z, then p, the frequency of A in the population, will be $\dfrac{2X + Y}{2(X + Y + Z)}$, because the homozygote AA carries two A's and the heterozygote one, and the total number of genes present is twice the number of individuals, for each carries two allelomorphs. Similarly q, the frequency of a, is $\dfrac{2Z + Y}{2(X + Y + Z)}$. A table can be constructed giving the possible types of mating, the frequency of each type and the offspring they will produce from Mendel's law of segregation. The frequency of the different matings can be obtained by multiplying together the frequencies of the genotypes as shown in the table.

	X (*AA*)	Y (*Aa*)	Z (*aa*) males
X (*AA*)	X²	XY	XZ
Y (*Aa*)	XY	Y²	YZ
Z (*aa*) females	XZ	YZ	Z²

Possible matings	*Frequency of mating*	*Proportion of offspring from each type of mating* AA	Aa	aa
$AA \times AA$	X²	X²		
$AA \times Aa$	2XY	XY	XY	
$AA \times aa$	2XZ		2XZ	
$Aa \times Aa$	Y²	$\frac{1}{4}$Y²	$\frac{1}{2}$Y²	$\frac{1}{4}$Y²
$Aa \times aa$	2YZ		YZ	YZ
$aa \times aa$	Z²			Z²
Total	1	$(X + \frac{1}{2}Y)^2$	$2(X + \frac{1}{2}Y)(Z + \frac{1}{2}Y)$	$(Z + \frac{1}{2}Y)^2$
i.e.		p^2 :	$2pq$:	q^2

From the above table it will be seen that the ratio of the genotypes $AA : Aa : aa$ in the first generation, whatever their frequency in the initial state, will be $p^2 : 2pq : q^2$. If we now calculate
the second generation using our new frequencies for the three
genotypes we obtain the following table:

Possible matings	*Frequency of mating*	*Proportion of offspring from each type of mating* AA	Aa	aa
$AA \times AA$	p^4	p^4		
$AA \times Aa$	$4p^3q$	$2p^3q$	$2p^3q$	
$AA \times aa$	$2p^2q^2$		$2p^2q^2$	
$Aa \times Aa$	$4p^2q^2$	p^2q^2	$2p^2q^2$	p^2q^2
$Aa \times aa$	$4pq^3$		$2pq^3$	$2pq^3$
$aa \times aa$	q^4			q^4
Total	1	$p^2(p^2+2pq+q^2)$	$2pq(p^2+2pq+q^2)$	$q^2(p^2+2pq+q^2)$
i.e.		p^2 :	$2pq$:	q^2

In the second generation, not only does the ratio of $p : q$ remain
the same, but also there is no change in the frequency of the
genotypes unless some disturbance such as natural selection
upsets the equilibrium. If the gene is sex-linked (in the X chromo-

some), equilibrium is not reached in one generation but will be reached in time, and the final steady frequencies will be achieved when the ratio of $AA : Aa : aa$ in the homogametic sex is $p^2 : 2pq : q^2$, and the frequency $A : a$ in the heterogametic sex (where, of course, only one gene will be present in each individual) is $p : q$.

This difference between the heterogametic and homogametic sex is the reason why a sex-linked recessive character will be more commonly found in the heterogametic sex. For example if the proportion of red-green colour-blind men in a human population is $\frac{1}{12}$ (q), that in women will be $\frac{1}{144}$ (q^2). Consequently, sex-linked recessives can sometimes be detected in a wild population, with a fair degree of certainty, without the necessity of doing breeding work.

IV

POLYMORPHISM

E. B. FORD has defined *polymorphism* as the occurrence together in the same habitat at the same time of two or more distinct forms of a species in such proportions that the rarest of them cannot be maintained merely by recurrent mutation.[26, 28] The phenomenon usually advertises a situation of particular genetic and evolutionary interest, for the presence of two or more distinct forms means that there is a balance of selective forces maintaining all of them in the population. R. A. Fisher has shown that if this were not so, one of the varieties would increase in frequency to the exclusion of the others, because it is very unlikely, except in unusual circumstances, that two varieties which are very different should be equally well fitted to the environment in which they both live.[20] Consequently polymorphism repays close study.

Before going on to describe how it can be maintained, it is necessary to discuss some of the limitations of the definition given above. This, like most biological definitions, is imperfect in the sense that it is not always possible to determine whether a particular example of variability constitutes a polymorphism. But in most instances the difficulties are more potential than actual and provided they are realized no real confusion can arise.

(1) The first point that has to be considered is what constitutes a distinct form. For example, it is clear that variation in human height does not constitute a polymorphism, as all intermediate heights between very tall and very short are found. In a species like the snail *Cepaea nemoralis*, on the other hand, most shells can be definitely classified as either yellow, pink or brown. In very large samples, however, there are occasionally just a few individuals which cannot be so easily classified, being intermediate in colour, either between yellow and pink or between pink and brown. Because the intermediate forms are so rare few

60

people would disagree that the colour classes in this snail constitute a polymorphism. However, from the theoretical point of
view, it does indicate a difficulty for it introduces a subjective
judgment as to how rare intermediate forms must be for variability to constitute a polymorphism. Fortunately in the majority
of cases the distinct varieties are controlled by two or more
allelomorphs of a major gene (a gene having an easily detectable
effect) which can be identified by observing segregation in
families so that no difficulty is encountered in classifying the
various types correctly.

(2) The control of polymorphism by major genes introduces
yet another difficulty because their identification depends on the
sensitivity of the method used for detecting the presence of the
various genes as to whether the population falls into distinct
classes or not. In many species of *Drosophila* there are differences
between individuals with respect to their chromosomes. The
differences frequently involve inversions (p. 53). Now inversions
can be clearly seen in the giant chromosomes found in the cells of
the salivary glands, and some other tissues of many two-winged
flies (Diptera) after they have been treated with certain stains.[73]
The presence of two different kinds of chromosome, one with an
inverted segment and one without, clearly constitutes a polymorphism in a population, for there will be three types of animal, two
homozygous (one for the inverted sequence, one for the unin-
verted sequence) and the heterozygote with both inversions
(Fig. 2). The rearrangements frequently affect fertility or
viability but not in a clear-cut way, and if we study these attributes, rather than the morphological changes in the chromosomes, we will get continuous variability and no distinct
classes. It is therefore necessary when describing a polymorphism
to state the type of variability and thus indicate the method by
which it was recognized. In *Drosophila* we would talk about
chromosomal polymorphism, in *Cepaea nemoralis* colour and
banding-pattern polymorphism, and in the case of the blood
groups of man, blood-group polymorphism.

(3) The definition, by using the words 'in the same habitat at
the same time', is designed to exclude from polymorphism
differences between geographical races and seasonal forms.
Geographical races can, however, be found together in the same

place at the same time because of migration, as, for example, in the case of people of very dark and very light skin colour who can be found together in many places of the world. In butterflies which have two distinct seasonal forms these are not usually found together. However, in exceptional circumstances they may be, if one of the forms lives long enough to be still present when the other one starts to emerge. Although these cases may come under the definition on a strictly literal interpretation, obviously they should not be called polymorphisms.

(4) The other major difficulty which arises is in deciding when a variety is and when it is not maintained by mutation alone. If there are two allelomorphs, A and a, which produce different effects but in which neither is better fitted to the environment than the other, there will be no selection in favour of one of the allelomorphs. Now, if a population exists all of whose members are homozygous AA, after a period of time a very small percentage of these genes will have changed spontaneously (mutated) to a. Consequently in the course of time the frequency of a in the population will increase. However, a will not completely replace A for there will usually be mutation from a to A so that a balance will be achieved, when the number of genes changing from A to a in each generation is counterbalanced by the number changing from a to A. If the mutation-rates are the same in both directions, the equilibrium will be reached when there are equal numbers of A and a present in the population. If A changes to a more often, then the balance will be reached when there are more of a than of A, the exact frequencies depending on the relative mutation-rates.

Such a situation would appear to be an example of polymorphism, but in fact it would not, because the rarer form is being maintained solely by mutation. Therefore, with the definition worded as it is, one might think that it is impossible to tell when discontinuous variability in a population constitutes a polymorphism and when it does not. Differences sufficiently distinct to allow the organisms to be classified into two or more groups can, however, seldom be neutral or near neutral with respect to their effect on survival, so that the rarer form cannot constitute more than 1% or 2% of the population if it is maintained by mutation alone. In practice the problem of deciding if a population is polymorphic or not hardly ever presents any difficulty. The maintenance of more than one form by mutation alone is discussed in more detail in Chapter VII, p. 117.

Environmental variation

Not all polymorphism results from the presence in the population of two or more types which differ by major genes, although it frequently is produced in this way. It is therefore desirable to discuss environmentally controlled discontinuous variation before considering those instances where it is controlled genetically. A particular gene will not necessarily produce exactly the same effect in all conditions, for the expression of the characters controlled by it will be modified not only by other genes present in the organism but also by the environment. Thus in man there is probably at least one major gene which converts brown eyes to blue, which are recessive. There are, however, also others which modify this effect, so that homozygotes for the recessive may have eyes ranging from light brown to the deepest blue. Consequently, although two blue-eyed parents will usually produce blue-eyed children, they may sometimes have offspring which have hazel or light brown eyes.

The environment can also act in a similar way. Identical twins are in fact identical with respect to their genetic constitution, but because they have not been subjected to identical environments during their development and growth they usually differ in some respects and can be told apart. Stocks of mice resulting from many generations of brother-sister (called *sib*) matings are usually genetically almost identical (Chapter VII, p. 116, explains the reason for this). Those who have examined such stocks which are piebald will notice that the patterns are not identical from mouse to mouse, and that therefore the differences must, in the main, be due to the effect of the environment alone.

Now although conditions vary from place to place and time to time, a polymorphism is not likely to result from environmental variation. It would be necessary for there to be two distinct types of environment, with no intermediate, present in the same place and time for a polymorphism to result, and this is an unlikely situation. In exceptional circumstances, however, these conditions may be fulfilled. When an environmentally controlled polymorphism does occur, it nearly always results from evolution controlled by rather special selective forces. For example, various mammals in the northern hemisphere are brown or grey in summer, but turn white in winter. This allows them to be inconspicuous to their enemies in summer as well as in winter, when a brown coat would be very conspicuous against a background of ice and snow. Natural selection has resulted in the evolution of a genetic constitution which produces a dark pelage in summer and under winter conditions a white one, with only a very short transitional period. Intermediates are not usually found because such a colour would not be inconspicuous against a brown background or against a white one. Thus the animal has two alternative and distinct environments (in this instance at different times of the year), one brown and one white, with an intermediate situation rare or of short duration. Natural selection has built up a genetic constitution which maintains a constant coat-colour over a wide range of environments, but at a critical stage there is an abrupt switch to a different colour. By having such a switch mechanism, the disadvantageous intermediate fur-colour is eliminated.

It is true that in this instance the two types of environment are usually found at different times of the year so that there will not normally be a polymorphism resulting; or at least only a temporary one during the period when some animals have changed and some have not. However, in an area where only some of the animals change in winter, there may be a seasonal polymorphism.

When there are two distinct environments present at the same place and at the same time, it is possible to get an environmentally controlled polymorphism. In these circumstances there will often be what K. Mather calls disruptive selection; that is, selection favouring two or more very different forms, but discriminating against intermediates between them. In other words, there are two or more distinct optimal expressions of the characters concerned, any of which are favourable (though not necessarily equally so), with a large number of intermediate ones which are disadvantageous. In many parts of the world there are two distinct background colours against which animals have to conceal themselves, one brown, the other green. Large animals often make themselves inconspicuous under such conditions by breaking up their pattern with bands, stripes, spots or by countershading (*see* Chapter IX, p. 146). However, small terrestrial animals, relying on their colour for concealment, can often hide themselves more effectively by being either green or brown, depending on the particular colour of the background on which they rest. Consequently, polymorphism for brown *versus* green forms is common amongst the larger insects, particularly in those stages of their life cycle when they are not very mobile, as for example when they are pupae.

Most Lepidoptera which pupate in the ground, in wood, in a tough opaque cocoon, or in some other concealed situation, are smooth and dark brown or black in colour. A large proportion of those which pupate in exposed positions are rough in outline and have colour patterns on them which make them harmonize with their background. This indicates that natural selection is effective in controlling their appearance.

Among the Swallowtail butterflies there are some, for example the mimetic butterfly *Papilio dardanus* Brown (*see* Chapter X), which always pupate under a green leaf, and the chrysalis is flattened and green so that it is extremely inconspicuous. The Swallowtail *P. glaucus* L., which does not normally pupate on leaves but on the trunks of trees, nearly always has a brown pupa. However, many other members of this genus, for example the European Swallowtail, *Papilio machaon* L., or the black American Swallowtail, *P. polyxenes* F., sometimes pupate on green leaves or stems, and sometimes on brown ones. In these species there are two distinct colour forms, one green, the other brown, but almost no intermediates are formed. As both are found in the same environment at the same time, this variation constitutes a polymorphism.

E. B. Poulton investigated the control of this difference between brown and green pupae in the butterfly *P. machaon* and came to the conclusion, in 1899, that the colour of the background determined what colour the chrysalis would be. Dr. C. A. Clarke and I have recently reinvestigated the situation and found that pupae on the brown stems of plants tend to be brown more often than green. Those on the green leaves, on the other hand, tend usually to be green. In one experiment, of 20 pupae, 7 which were not on green leaves were brown, whereas only 1 was green. All of the 12 which were on the green leaves were green. A simple statistical calculation shows that the probability of this correlation being due to chance is far less than 1 in 1000; thus the experiments confirm Poulton's findings. However, experiments which are still in progress indicate that there are factors other than the colour of the background which are important in determining which form of pupa is produced, but just how the various factors interact is not yet understood. Nevertheless, it is clear that natural selection has evolved a genetic constitution which tends to produce green pupae in green surroundings and brown ones in brown. Moreover, the shade of the brown often harmonizes with the exact shade of the background.

There is evidence that different races respond differently to the environment (light, temperature, humidity, etc.), so that one race may produce mostly green pupae under conditions in which the other will produce mostly brown ones. It therefore appears as if the response of the organism to its environment is genetically controlled. This result would be expected, for selection will have adjusted different races to the particular conditions each is likely to encounter, to produce the appropriate colour. These will vary from race to race because the range of physical conditions over which it will be advantageous to produce green pupae will be different from place to place.

In areas where there are marked seasona¹ changes in vegetation it is not sufficient for a green pupa to be produced on a green background for this is likely to change to a brown one in some seasons, thus making the pupa conspicuous. For the polymorphism to be effective there must be some mechanism which ensures that under normal conditions a green pupa, for example, does not remain in full view of predators throughout the winter. In other words, one would not expect a green pupa to go into a state of hibernation as often as a brown one, but to emerge as a butterfly in the late summer. Now many of the species with the two types of pupae are double- or triple-brooded with only a proportion of the pupae emerging to form the last brood, the others remaining as chrysalises through the winter. Consequently the winter population of insects consists of these individuals and the offspring of their brothers and sisters which formed the last partial brood. One would expect that there would be no difference in the speed of development of the two kinds of chrysalids in the early summer. However, because green is a disadvantageous colour in winter one would expect that, in the latter part of the year, a greater proportion of the green pupae would tend to emerge compared with the brown ones.

In September 1955 I received from America 27 pupae of *Papilio polyxenes*: 12 of these were green and 15 brown. All the green ones produced butterflies late in the same month, whereas only two of the brown ones did so. The remaining pupae over-wintered and emerged as butterflies in the spring. The probability of such a result being due to chance is remote. It therefore seems that, as would be expected on the argument put forward above, green pupae tend not to hibernate. Exactly the same situation has been found by S. R. Bowden in the white butterfly *Pieris napi*, which has two kinds of chrysalis, and I too have observed it in this species.

The experimental and observational data suggest that natural selection has built up a genetic constitution in these and probably other species which enables the organism to react appropriately to a complicated set of environmental factors. The result of this is that the chances of survival of the individual are enhanced. It seems obvious that environmentally controlled polymorphisms deserve far greater attention than has been given them so far.

Transient polymorphism

Genetically determined polymorphism, which J. S. Huxley has named *morphism*,[42] arises when alternative forms (allelomorphs) of major genes are common in a population. There are at least three possibilities which would explain the presence of this diversity.

(i) The various forms may all have the same selective value, so that the gene frequencies remain constant from generation to generation (p. 57).

(ii) There may be a balance of selective agencies so that no form tends to be eliminated. This will happen if the rarest allelomorph is always the advantageous one, its advantage decreasing and being converted into a disadvantage as it becomes commoner. Polymorphisms maintained in this way are called *balanced* or *stable polymorphisms* (*see* Chapter V).

(iii) A new mutant may be advantageous when it arises and spreads through the population. Alternatively, a previously rare form may become advantageous and spread throughout the population because the environment has changed, converting its disadvantage to an advantage in comparison with the other allelomorph. During the period when it is spreading there will be a polymorphism. This situation, is, however, transient because the polymorphism will disappear when the new advantageous form has replaced the original one, or reduced it to such a low frequency in the population that it is only retained as the result of recurrent mutation (*see* Chapter VII, p. 117). Such polymorphisms are called *transient* because of their fleeting nature.

R. A. Fisher[20] in 1930 pointed out that the range of conditions over which a gene could be neutral in selective value as compared with its allelomorph must be very small indeed. Consequently, as the environment is by no means constant, genes can seldom be neutral in effect for more than a very short period of time. Moreover, the larger the effect of the gene the less likely it is to be neutral, so that discontinuous variability maintained as a result of the allelomorphs having the same selective value must be so rare that it can be ignored in the present discussion, although the matter is dealt with briefly in Chapter VII.

Stable polymorphisms are probably commoner than any other since they will persist while conditions remain approximately the same and often even under considerably different ones, so that examples of them will accumulate. Because they are so common, and have been investigated in much detail, they will be dealt with at more length in the next chapter. Transient polymorphisms are less common because they will only be present in a population for a short period of time, biologically speaking, and it is these which will now be considered.

The most spectacular evolutionary change ever witnessed and recorded by man concerns a number of examples of transient as well as of some stable polymorphism. This is the phenomenon of *industrial melanism* and, in England alone, as many as seventy species of moths are altering their appearance as the result of a change in the countryside caused by the spread of industrialization.[24], [45] On the continent of Europe, particularly in Germany, the same evolutionary change has been observed, and there is now some evidence that this is also taking place, though to a less marked degree, in the United States.

The Industrial Revolution has profoundly altered the countryside in many ways. One of the important causes of this change is the vast amount of soot and other waste products being poured out from factory and dwelling-house chimneys which pollutes the countryside for miles around. It is mainly this pollution that has caused the evolution to be described. Before discussing this, it is convenient to give a brief history of the changes which have occurred in one species, the Peppered Moth, *Biston betularia* (L.). (The following account is based on the earlier work of E. B. Ford and on the recent extensive field investigations of H. B. D. Kettlewell.)

Prior to the middle of the nineteenth century the insect, so far as is known, was always white with black specklings on the wings and body (Fig. 3a). This will be referred to as the typical form. In Manchester in about 1850 a black variety of the species was caught for the first time (Fig. 3d), and this is now called *carbonaria*. The form, which is controlled by a single dominant gene, appears not to be so black in some of the early specimens as it now is even in heterozygotes (Fig. 3c), a circumstance that will be referred to in Chapter VIII. *Carbonaria* increased steadily in frequency from the middle of the nineteenth century until the present day. Moreover, not only did it become extremely common in Manchester, but it appeared also in other manufacturing districts until, at the present day, it is common in all of them, often reaching a frequency of 95% or more. Consequently, although about a hundred years ago the typical form was the only one in most manufacturing districts, today it is a rarity in them. This, however, is not the whole story, for *carbonaria* is also spreading in many apparently rural areas far from industrial cities, although it has done so more commonly to the east than to the west of such cities. It will be seen from the map (Fig. 4) that the only places in the British Isles which are apparently free of these black varieties are Northern Scotland, West Devon and Cornwall, parts of the South Coast of England, particularly in the

FIG. 3

There are three main forms of the black and white Peppered Moth, *Biston betularia*. (a) The typical form and the two dominant melanics (b) *insularia* and (c) and (d) *carbonaria*. The two specimens of *carbonaria* are not identical; (c) is a heterozygote (black being dominant) and (d) is a moth caught in the middle of the nineteenth century and probably a heterozygote. The difference between the modern forms and some of the early melanics suggests that, at the present time, the allelomorph controlling form *carbonaria* produces a blacker insect than it did a century ago. The melanic *insularia* (b) although looking intermediate between typical and *carbonaria* is in fact controlled at a different locus.

West, and probably parts of Ireland and Wales. Although the Peppered Moth was the first to show this change, at the present time many other species are also doing the same thing, both in Great Britain and on the continent of Europe. Moreover, each year new black forms are being found in species in which they were not previously known, and nearly all so far investigated genetically have proved to be controlled by major genes which are dominant or have an intermediate heterozygote. [24], [26] None of the true industrial melanics have yet been shown to be recessive, despite the fact that equally black recessive forms are known to occur as rarities in some species.

70

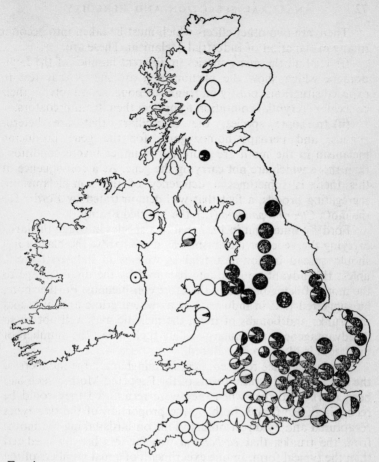

Fig. 4

Sketch-map showing the frequencies of the three forms of the Peppered Moth, *Biston betularia*, in Great Britain. The circles show the localities in which collections were made and their size indicates the number of moths in the sample. The largest circles represent collections of over 100 moths and the smallest samples of between 25 and 50 moths. The size of the black area of each circle gives the percentage of *carbonaria*, the hatched area the percentage of *insularia* and the white area the percentage of the typical form. A single dot in the centre of the circle means that only one moth of that variety was taken in the collection. Note that the proportion of *carbonaria* is high in industrial regions and to the east of them but decreases rapidly to the west, south or north. The proportion of *insularia* compared with typical behaves in much the same way, showing that it also is an industrial melanic.

There are two other effects which must be taken into account in any explanation of Industrial Melanism. These are:

(i) That all the many species in different families of the Lepidoptera which show the phenomenon are moths that rest in exposed situations, mostly on trunks or boughs, and rely on their concealing (cryptic) coloration to protect them from predators.

(ii) In some species, the caterpillars that are heterozygous, and perhaps homozygous, for the gene producing melanism in the moth are more hardy under severe conditions than those which are not carrying the gene. As a consequence of this there is sometimes a deficiency of the non-melanics in segregating broods, a fact that was demonstrated by Ford[25] for the moth *Cleora repandata* (L.), the Mottled Beauty.

Ford[24] pointed out in 1937 that in a species in which the larva carrying the gene for melanism was more hardy, the black form should spread in non-industrial as well as in industrial areas, unless this advantage is counterbalanced by the disadvantage to the moth of being black and therefore conspicuous. Furthermore, he suggested that in industrial areas where grime and filth coat the trunks and boughs of trees, the melanic may well not be so disadvantageous, particularly as in such areas the number of predators is likely to be reduced.

Kettlewell[45] has released, near Birmingham, large numbers of the black and the typical forms of the Peppered Moth, which had been marked so that when they were recaptured they could be recognized. He found, both by the proportion of the two types recaptured and by direct observations on birds taking the moths from the trunks, that *carbonaria* was far less heavily predated than the typical form. In one experiment, of equal numbers of the two forms 43 typical were taken to only 15 *carbonaria*. Consequently the melanic was at a great advantage, which explains why it has become so common in polluted areas. By repeating the same type of experiment in a non-polluted wood in Dorset he showed that in such an area this melanic is very much more conspicuous than the typical form, and is far more heavily predated, thus putting the typical form at an advantage. Five species of birds between them took 164 *carbonaria* but only 26 typicals when equal numbers of the two varieties were put on tree trunks. This, of course, explains at once why moths with the typical pattern are common in the unpolluted countryside but are rare in industrial areas.

The most important factors in determining which variety is the less readily seen are the presence or absence of conspicuous patches of lichens on the trees, and the amount of grime and filth. In areas where there is smoke pollution, even if it is not very great, these patches disappear or are very much reduced in number. The typical form of the Peppered Moth has a pattern which renders it very inconspicuous on lichened trunks or boughs, but when these are absent and the trees are black the pattern makes the moth far more conspicuous. In the same way, *carbonaria* is inconspicuous in the absence of lichens, and particularly when the bark is dark, but is very obvious in the presence of lichens. To the east side of industrial areas patches of lichens on trees tend to be rare or absent over great distances because smoke and pollution are carried a long way by the prevailing westerly winds. However, on the west side of industrial areas lichens disappear for a much smaller distance because less smoke reaches them. This explains why in the Peppered Moth melanics are common on the East Coast of England, even as far east as Norfolk, where they reach a frequency of 70% or 80%, but are rare or absent in the west, where lichens tend to be more in evidence (Fig. 4).

Although this account has given the salient features of the situation, the matter, as Kettlewell has shown, is not as simple as it might appear. For example, Kettlewell[44] has evidence that the gene responsible for the melanism may have other effects, particularly on the behaviour of the melanics. He has some data which suggest that they tend to rest on darker surfaces than the typical forms, and this would, of course, affect their conspicuousness, and therefore their selective value. Moreover, in the Peppered Moth there is another gene which produces a variety intermediate in appearance between *carbonaria* and the typical form. The variety produced is called *insularia* (Fig. 3b) and would appear at first sight to be the intermediate heterozygote between *carbonaria* and typical. This, however, is not so. The pattern is dominant or nearly dominant and is controlled by an allelomorph at an entirely different locus. Because it increases the amount of black on the animal, it has a distinct effect when the moth is not *carbonaria*, but when the insect is already black the allelomorph can have no visible effect. Consequently there are only three, not four, distinct types, the typical form, *insularia* and *carbonaria*. If it were possible to distinguish *insularia* in an animal which is *carbonaria*, there would have been a fourth phenotype. In the absence of this, it is only possible to discover by breeding experiments whether an individual which is *carbonaria* is also carrying the gene for *insularia*.

The distribution of *insularia* in the British Isles at once shows that it also is an industrial melanic, for the gene tends to be common in areas affected by industrialization and to the east of them, but absent or almost absent from other regions (Scotland, Wales, Ireland, West Devon and Cornwall). However, its distribution is different from that of *carbonaria* in that it is often at a high frequency in marginal regions (those little affected by smoke pollution) where *carbonaria* is still fairly uncommon. Thus it is at a higher frequency than the other melanic in the southern part of England and in the south-west, particularly the Gloucester and Severn Valley area as well as in the Isle of Man. This suggests that in a region which is beginning to be affected by pollution, the *insularia* pattern becomes advantageous before *carbonaria* does. This would be expected, for it is a less extreme melanic and therefore is likely to be at an advantage under less extreme conditions.

There is another interesting difference between the two forms. The gene producing *insularia* is never found in much more than 50% of the individuals in a population, whereas that for *carbonaria* is frequently found in over 90% in some areas (Fig. 4). There seem to be two possible interpretations of this fact:

(i) It might be argued that, with increasing industrialization, conditions usually become extreme in a short period of time. Consequently *carbonaria* increases in frequency, and *insularia* loses much of its advantage, because few individuals will be of this form, most being *carbonaria*. With this loss of advantage the gene will cease to spread rapidly and will remain for a long period at about the same frequency or decline slowly.

(ii) If the homozygote *insularia* is at a disadvantage compared with the heterozygote, then in these circumstances there will be a balance of selective forces which will allow *insularia* to increase up to the frequency at which its advantage in the heterozygote is balanced by its disadvantage in the homozygote. This situation will result in a stable polymorphism with both allelomorphs maintained in the population, a subject which is discussed in the next chapter.

It seems unlikely that the first explanation is the correct one, for various reasons which cannot be considered here. It is more probable that an increase in the frequency of *insularia* above a certain level is prevented by the homozygote being disadvantageous. This is likely to result from some physiological effect, for the heterozygotes and homozygotes are similar in appearance. There might well be a difference in viability or fertility between the two. This interpretation is further supported by the fact that other species of moths are known which are polymorphic for a melanic (including industrial melanics) and in which the frequency of the two forms remains stable. Moreover, the commonness of the allelomorph producing *insularia* even in areas where the variety can have little advantage from the point of view of its appearance, since most moths are *carbonaria*, suggests that the allelomorph may confer extra hardiness on the larva, as is the case in the Mottled Beauty[25] (p. 72).

In this section Industrial Melanism has been treated in considerable detail, not because it represents the only known example of transient polymorphism but because it illustrates the situation so well, and has been investigated in more detail than any other.

Most other examples of transient polymorphism concern genes which start to spread when conditions are favourable to them, but then decrease and disappear when the environment again becomes unfavourable. This is what would be expected. Most genes that are favourable to their carriers will have already replaced their allelomorphs so that unless there is a drastic change in the environment, no new genes are likely to be observed spreading in a population in the way those controlling Industrial Melanism have done. New advantageous allelomorphs would become common but the occurrence of these must be extremely rare in stable conditions, for, because mutation is a recurrent phenomenon, most of them will have occurred in the past. Genes are unlikely to be observed spreading because there must have been a change in the environment for them to do so; moreover, for their increase in frequency to be so rapid that it can be detected in the course of a few generations, the selective values involved must be very large. It is therefore not surprising that the best examples of transient polymorphism concern Industrial Melanism, because here a very profound change of the environment has occurred in the last 150 years, allowing us to observe evolution in progress.

V

STABLE POLYMORPHISM

As has been mentioned in the previous chapter, stable polymorphism results when there are opposing selective forces operating which ensure that two or more allelomorphs of one gene are maintained in the population. The forces which do this usually operate in such a way that an allelomorph is at an advantage when it is rare, but at a disadvantage when it is common. For in these conditions, as the gene starts to become rare it starts to become advantageous and there will be a tendency for it to increase in frequency. However, it will decrease when it is common and therefore disadvantageous. Consequently the situation is stable and gene frequencies will remain close to a value at which all allelomorphs have the same advantage. Several conditions can give rise to this state, and some of them will be considered below.

Stable polymorphisms are far commoner than transient ones because they tend to persist and their number to increase as new ones arise. Unstable polymorphisms, on the other hand, do not remain in existence for long, even under constant conditions; consequently few of them can accumulate in a population.

Some stable polymorphisms are known to have persisted for thousands of years. C. Diver[17] pointed out that the various banding forms in the snail *Cepaea* are found in fossils and sub-fossils dating back certainly to the Neolithic Age and probably further. Moreover, the fact that closely related species are sometimes polymorphic for very similar varieties is strong evidence that the polymorphism is older than the species themselves, and that their common ancestor was also polymorphic. This situation is particularly evident in *Cepaea nemoralis, C. hortensis* Müller and their relatives, which have a very large number of varieties in common, and in man in whom there are polymorphisms which

77

are also found in the great apes (some blood groups and the ability or inability to taste phenylthiourea).

Heterozygotes at an advantage to homozygotes

When there is, in a population, a pair of allelomorphs whose heterozygote, in the prevailing conditions, is at an advantage to both homozygotes, the two allelomorphs will be maintained in the population while conditions remain approximately the same. That this will happen can easily be seen in the extreme case where both homozygotes are lethal or sterile. In these circumstances a population composed entirely of heterozygotes *Aa* will produce the genotypes *AA*, *Aa* and *aa* in the ratio of 1 : 2 : 1 (*see* Chapter III, p. 57). However, the homozygotes (*AA* and *aa*) will die or be sterile so that the individuals which breed successfully in the next generation will all be of the constitution *Aa*, and this, of course, will be true for all subsequent generations if the environment remains unaltered. Consequently the two allelomorphs will be represented in the breeding population in equal numbers in each generation.

The two allelomorphs will be maintained in much less extreme conditions when both homozygotes are only slightly inferior to the heterozygote. If we take a population in which all the organisms are homozygous for a gene *B*, they will have the genetic constitution *BB*. Now if a new form *b* appears, individuals carrying it will seldom be homozygous when it is uncommon, because two will seldom meet in the same individual. The frequency of the three genotypes will be approximately $p^2 : 2pq : q^2$ (Chapter III). This means that the homozygote *bb* will be very rare indeed compared with the heterozygote when q (the frequency of *b*) is small. Consequently if *Bb* is even at a slight advantage to *BB* the frequency of *b* will increase because the heterozygotes will contribute proportionately more offspring to future generations than will the homozygote *BB*. This will be true even if *bb* is at a disadvantage to *BB*, for being rare the number of *b* genes lost to the population as the result of the disadvantage of the homozygote will be far outweighed by the gain due to the advantage of the very much commoner heterozygote. However, as *b* increases in frequency, the proportion of *bb* homozygotes compared with that of the heterozygotes will also increase. When the homozygote becomes sufficiently common, a stage will be reached at which as many *b* genes are lost from the population, because of the homozygote's disadvantage, as are gained as a result of the advantage of the heterozygote. At this stage a stable state will be attained.

A little further reflection will show that the frequency at which stability is reached will depend on the relative selective values of the two homozygotes. A mathematical proof that stability will result if the heterozygote is at an advantage to both homozygotes has been given by J. B. S. Haldane, R. A. Fisher, S. Wright and others.

Drosophila melanogaster can be used to illustrate how gene frequencies will change as a stable state is approached. A stock of normal *melanogaster* was kept in the laboratory and a certain number of individuals heterozygous for the mutant stubble (*Sb*) were introduced. The gene is lethal when homozygous and affects the length of the bristles on the thorax in the heterozygote. In this particular experiment just over 60% of all non-stubble flies (*sbsb*) were removed in each generation. Consequently the non-stubble homozygote *sbsb* was (artificially) at a disadvantage to the heterozygote *Sbsb*, as was the stubble homozygote *SbSb* on account of its lethality. The table gives the results of this experiment. As will be seen, the proportion of the *Sb* genes increased rapidly at first, but is finally stabilized at a frequency of about 0·365.

Generation	% Stubble	Frequency of Stubble Allelomorph
1	14·3	0·0715
2	33·7	0·1685
3	57·6	0·2880
4	63·2	0·3160
5	69·1	0·3455
6	73·5	0·3675
7	72·9	0·3645
8	73·4	0·3670
9	72·9	0·3645

Note that after generation 5 the gene frequency of stubble maintains an equilibrium value of about 0·365.

In this example the gene frequencies can be calculated, because it is known that the stubble homozygotes do not survive, so that all stubble individuals must be heterozygotes. The frequency of the allelomorph *Sb* equals the number of stubble genes divided by the total number of genes at that locus in the population. Consequently, in this example, the number of stubble genes will be the number of heterozygotes (each carrying one), and the number of genes at that locus in the population will be twice the number of flies, for each locus is represented twice in every individual. In general, the frequency of a gene *A* in a population will be twice the number of *AA* homozygotes plus the number of heterozygotes (*Aa*), divided by twice the total number of individuals (*see* Chapter III, p. 57).

When the three genotypes can be recognized, it is possible to calculate their relative selective values in a population if their proportions are changing from generation to generation. When there is random mating and no selection, the frequencies of the three genotypes will be $p^2 : 2pq : q^2$ (if the gene is not sex-linked). However, if there is selection, then in the next generation their proportions will be $ap^2 : 2bpq : cq^2$, where a, b and c are the relative selective values of the three genotypes. In the present example, *SbSb* is lethal, so that if p represents the frequency of *Sb* the value of a will be 0. Consequently, if we take the value of b to be 1, the value of c can be calculated. This is particularly easy when the final equilibrium frequency is known, for the frequency at which this point is reached is given by the formula $\dfrac{1-c}{(1-a)+(1-c)} = p$. Therefore in this example $\dfrac{1-c}{1+(1-c)}$ is approximately 0·365. Therefore $c = \dfrac{0·270}{0·635} = 0·425$. Because about 60% of the *sbsb* individuals were removed in each generation, we know the value of c, which will be about 0·4. This corresponds well with the value of 0·425 calculated from the equilibrium frequency.

Armed with this value of 0·425, we can calculate the expected gene frequency in each generation from the beginning of the experiment. As might be expected, if one does calculate these values one finds that they do not agree *exactly* with the observed values. This could be due to chance alone, resulting from the fact that the population was not very large (*see* Chapter VII), or due to conditions changing during the experiment and thus altering the relative selective values of *Sbsb* and *sbsb*.

In nature environments are anything but constant, varying from day to day, season to season, and year to year. Cyclical changes, such as those which occur within the period of a day, will not cause a corresponding change in gene frequencies in most organisms (with the possible exception of some bacteria and viruses), for the generation time is longer than one day. Changes during the year, however, may cause rhythmical fluctuations in gene frequency. For example, N. W. Timoféeff-Ressovsky[67] investigated a beetle, the Twin Spot Ladybird (*Adalia bipunctata* (L.) near Berlin. He found that the frequency of the red forms increased during the winter when the beetles were in hibernation, but that the black forms became commoner in the summer. By comparing the ratio of the two forms in the living and dead beetles during the winter he showed that the black form did not survive so well as the red one in these conditions. The increase in black during the summer shows that it is at an advantage to the red form during this season.

Dobzhansky found a rather similar situation in the fruit-fly *Drosophila pseudoobscura*. This is one of the insects in which certain cells have very elongated and enlarged chromosomes. These are particularly evident in the salivary glands, where they are so enlarged that they can easily be seen under the microscope at low magnifications. Now in this species there are a number of inversions present (Chapter III, p. 53). They can be detected because each pair of chromosomes has a characteristic pattern of bands in it, and if an inversion is present in a particular chromosome the order of the bands is reversed over the part containing the inversion. Moreover, in a heterozygote the detection is even easier than in a homozygote. The genes in one of a pair of chromosomes tend to lie next to their allelomorphs in the other and consequently, when their order is reversed in one, both tend to become looped, so that similar loci still lie next to one another (Fig. 2). This looped configuration is quite characteristic of inversion heterozygotes and can readily be recognized in suitably stained preparations of larval salivary glands.

Three of the inversions studied by Th. Dobzhansky[18] are called

Standard (ST), Arrowhead (AR), and Chiricahua (CH). He found by taking collections from a wild population that the frequency of ST was high in the spring, dropped steadily until June and then rose gradually to a high level in October. CH on the other hand was at a low frequency in March, rose steadily in frequency until June, and then fell steadily until October. The frequency of AR also altered to some extent, but not so markedly as did the other two arrangements. These changes indicate that STST is at a disadvantage compared with CHCH from March to June, but at an advantage from June to October, there being apparently little change in the frequencies of the three types during the winter. By keeping populations of flies in wooden boxes (called population cages), he demonstrated that at a temperature of about 25°C. ST will reach a stable equilibrium at a frequency of approximately 70% if it is present in a population cage with CH. At 16°C., on the other hand, the forms remain at the frequency at which they were in the population when it was started. This shows that at 16°C. there is no, or very little, selection operating on the effects produced by these inversions.

The results show why ST increases during the warm part of the year from June to October, but they do not show why the ST homozygote is at a disadvantage to the CH homozygote in the spring. Recently L. C. Birch[3] found that when the larvae were crowded, as they are in a population cage, and probably are in the wild in summer, the ST homozygote was at an advantage to the CH homozygote. But he also found that when the larvae were not crowded the CH homozygote was at an advantage to the ST one, so that instead of ST reaching a stable equilibrium when it was at 70%, CH reached this frequency and ST was stable at 30%.

Now it seems very likely that the larvae are not crowded in the wild in the spring, for at this period the population is sparse, whereas it becomes dense in the summer. Consequently Birch's results probably explain why the frequency of CH increases from March to June. It is important, however, to stress that this interpretation of the facts is by no means proved, for the breeding sites of *Drosophila* in the wild are still very imperfectly known. What is proved beyond doubt is that selection is acting and the seasonal changes result from the changing selective values of the homozygotes under different conditions.

The reason for the commonness of chromosomal inversions in many species of animal and plant is not fully understood. However, they are known partially to suppress crossing-over between genes within them because of mechanical difficulties caused by the loop formation when the chromosomes pair at cell division (p. 53). Moreover, when a break does occur and crossing-over takes place, the new chromosomes tend not to survive, because one of them will have some of the loci represented twice, whereas in the others these will be missing. Both of these new chromosomes will usually be lethal to their carriers. Consequently if an inversion arises which includes a combination of genes which is advantageous, the inversion will tend to survive because it will maintain the combination and will prevent the genes reassorting themselves into less advantageous combinations as a result of crossing-over.

The advantage of certain combinations of genes over others is fully discussed in Chapter VI, but the principle can also be illustrated here by some data from the American Grouse Locusts *Paratettix texanus* Hancock and *Apotettix eurycephalus* Hancock. In these species there is a common inconspicuous form. There are also a number of much less common and very distinct varieties present in most populations. Fisher showed that, in R. K. Nabours's extensive breeding data for *A. eurycephalus*, the heterozygotes for each of these varieties were at about a 7% advantage over the dominant homozygote. He also found that they were at about a 5% or 6% advantage over the common inconspicuous form, which is the recessive. Such a condition, if present in the wild, will ensure a stable polymorphism (p. 78).

It seems very probable that the same situation holds for *Paratettix texanus*. However, from collections taken from wild populations, Fisher[21] showed that there was a deficiency of individuals heterozygous for two or more dominant genes. This could only be explained if it were assumed that a single heterozygote was at an advantage to both homozygotes, but that an individual which was heterozygous at two loci was at about a 40% disadvantage. Thus although the genotype *Aabb* was at an advantage to *aabb* and *AAbb*, and *aaBb* was at an advantage to *aabb* and *aaBB*, the genotype *AaBb* or any other genotype manifesting two dominant characters was at a very great disadvantage to all the others. In these circumstances one would expect very strong selection in favour of any genes suppressing crossing-over between any two of these loci, for in an individual with the chromosome constitution $\dfrac{A\ b}{a\ B}$ a cross-over occurring between the loci would produce the chromosomes $a\ b$ and $A\ B$. The latter would be severely handicapped, for any individual carrying it would be at a 40% disadvantage to most other genotypes. It is perfectly true that a cross-over in a double heterozygote of the constitution $\dfrac{A\ B}{a\ b}$ would be advantageous, since it would produce the chromosomes $A\ b$ and $a\ B$, but double heterozygotes of this type will be much rarer than the other because the chromosome $A\ B$, being very disadvantageous, will be rare in the population. As a result of the selection, any inversion including the two loci and including the combination Ab or aB will reduce the chance of the combination AB being formed by crossing-over, and will therefore be at an advantage in this respect. In fact, the genes controlling the dominant varieties in these two species of Grouse Locusts and in several other species are very closely linked, crossing-over between them being rare. It seems probable that this close linkage was not present originally, but has been evolved as the result of natural selection for the suppression of crossing-over between the loci involved.[62] This could either be achieved by selecting inversions or alternatively genes which reduce or suppress crossing-over in the appropriate parts of the chromosomes. Such genes which affect the frequency of crossing-over have been noted in several organisms.

The presence of close linkage between genes which contribute to a polymorphism is of widespread occurrence,[62] and is found in many groups of animals and plants, including the land snails *Cepaea nemoralis* and *C. hortensis*. Both these snails show considerable variation in colour of the shell and the pattern of bands on it, and most of the varieties are found in both species. Populations of *C. nemoralis* tend to be more variable than those of *C. hortensis*. The colour of the shell may be yellow, pink or brown, and the commonest patterns are unbanded, 5-banded and 1-banded (Fig. 5). Brown is dominant to the absence o

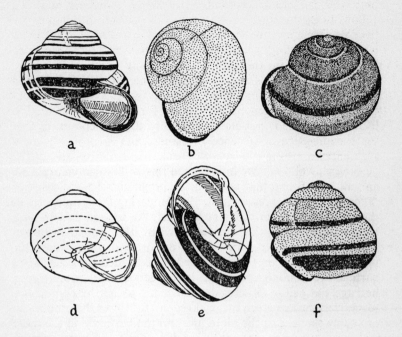

a

b

c

d

e

f

FIG. 5

The land snail *Cepaea nemoralis* is highly polymorphic and a few of the known varieties are illustrated: (a) a yellow shell with five bands and a dark lip at the mouth of the shell; (b) a pink shell with no bands and a dark lip; (c) a brown shell with only the central band present; (d) a yellow shell with the bands present but unpigmented, making them translucent (the lip is also unpigmented); (e) a yellow shell with pigmented bands but an unpigmented lip; (f) a pink shell with the first two bands missing so that it has only the central and two lower bands present. All combinations of shell colour and banding pattern can be found but some are much commoner than others. Brown is dominant to pink which is dominant to yellow, the three apparently being controlled by three allelomorphs. Unbanded is dominant to the presence of bands and the locus is linked to that controlling colour. The one banded form (c) is dominant to five banded and is not controlled by the locus which determines the presence or absence of bands. The form with unpigmented bands and a white lip is recessive to varieties with pigment present in these areas. The absence of the two upper bands is probably dominant to their presence.

brown, and pink is dominant to yellow. The gene controlling the presence of brown is very closely linked to, or perhaps allelomorphic with, the gene determining the difference between pink and yellow. In the banding series the absence of bands is dominant to their presence, and the locus is very closely linked with the locus or loci determining colour. If bands are present, there are often as many as five on the shell. However, any single band or any combination of bands may be missing. One particularly common form has only the third band present, and this variety is controlled by a single gene, the 5-banded form being recessive. The locus responsible is not closely linked to that for colour, or that for the presence or absence of bands, and is probably on a different chromosome.[8]

Several workers have investigated this striking polymorphism, and all have found that nearly every population has at least two different varieties present in it. By comparing the frequencies of the various forms in different habitats, A. J. Cain and I[7] have been able to show that in woods where the ground is brown with decaying leaves, and fairly uniform in appearance, the unbanded brown or the unbanded or 1-banded pink shells are particularly common, whereas in hedgerows and rough green herbage the yellow 5-banded form tends to be the commoner. Moreover, the less green the background the lower the proportion of yellow shells, and the more uniform the background the more common the unbanded or 1-banded forms. This association between habitat and the frequency of the varieties can be clearly seen in the diagram (Fig. 6). Along one axis has been plotted the proportion of yellow snails, along the other the proportion of effectively unbanded ones. This class comprises all shells with bands 1 and 2 absent, including, of course, the totally unbanded shells. This classification has been adopted because bands 1 and 2 are the more conspicuous in the living animal and when they are absent the shell tends to look much less banded than when they are present.

The Song Thrush, *Turdus ericetorum* Turton, breaks open these shells on suitable stones and leaves the broken remains behind. Consequently one is able to compare the proportion of the varieties in a colony with those taken from it and eaten by thrushes. In this way it has been possible to show that in certain rough habitats

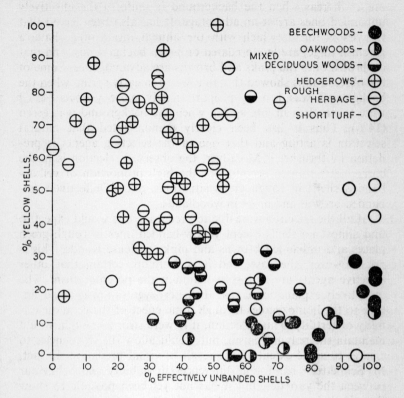

FIG. 6

Correlation-diagram for per cent yellow shells, per cent effectively unbanded shells (those with at least the upper two bands absent Fig. 5, b, c and f) and habitat of colonies of *Cepaea nemoralis*. The diagram shows the way in which natural selection has affected the frequency of the colour and banding patterns. Thus in woods, which tend to have a uniform carpet of brown leaves (particularly beech woods), the proportion of the less conspicuous effectively unbanded brown or pink shells is high. In contrast to this, habitats with rough herbage (including hedgerows) have the proportion of yellow banded shells high as these shells are less conspicuous in such places. The few exceptional colonies are also in accord with this view; thus the mixed deciduous wood population with 80 per cent yellow shells lived in an exceptional wood in that the ground was carpeted with short green grass and not brown leaves, hence the high proportion of yellow effectively unbanded shells.

the banded forms are at an advantage to the effectively unbanded ones, whereas when the background is uniform the effectively unbanded ones are at an advantage. It has also been found that the yellow forms (which with the animal inside often have a greenish tinge) are less predated on green backgrounds, and that on brown ones the pinks and browns are advantageous. Some of the observations showed that in a wood in early spring when the background was brown, proportionately more yellows (43%) were taken than in late spring when the background was green (14%). Thus it has been clearly demonstrated that natural selection is acting and that one of the selecting agents is predation by thrushes.[59] Moreover, the observed selection is in the correct direction to account for the high proportion of yellow-banded shells in rough green situations, and of effectively unbanded browns and pinks in woodlands.

If all the selection were due to predators, one would expect to find almost no shells except yellow-banded ones in rough green places and unbanded browns and pinks in dense woods. This is not, however, the case, and it is therefore certain that other selective agents are acting to maintain the polymorphism. The most likely explanation is that the heterozygotes are at an advantage to the homozygotes through some effect of the gene on viability or fertility. This selection, if it were strong enough, would maintain the polymorphism, but would allow the frequencies to change to some extent as the result of visual predation. In fact, M. Schnetter[57] has found differences in viability and behaviour between the varieties, but it has not yet been possible to show that the heterozygotes are at an advantage in respect of these differences.

It is in man that we have some of the most complete evidence on how a physiological difference between the heterozygotes and the homozygotes can maintain a polymorphism. The red blood pigment which transports oxygen in the body is called haemoglobin, and in the foetus most of it is different in constitution from that found in individuals of over a year old. We can for convenience call the red pigment in the foetus haemoglobin F, and that in children and adults A. Now in certain parts of the world, notably Africa, there is a third type, S, whose presence is controlled by a single gene which we can represent by Hb^S, there being no dominance, and the normal allelomorph, producing A haemo-globin, by Hb^A. In the heterozygote $Hb^A Hb^S$, the red cells look normal. However, if treated in a special way they become distorted and are then said to 'sickle'. This does not happen in the body because the amount of oxygen present has to be reduced below that normally found in the living human being for it to occur. In the heterozygote there is usually rather less S than A haemoglobin, but the difference in the amount of the two types is not often great. However, in the homozygote $Hb^S Hb^S$, there is no A red pigment, all of it being S, except for a small quantity of F which is not normally present in adults. Because of the total absence of normal red pigment the cells can sickle even in the living body. They then tend to block the blood vessels and to be destroyed by the body, thus causing severe anaemia. This often proves fatal.

A. C. Allison[1] has shown that in parts of East Africa, where the Hb^S gene reaches a frequency of 20% or more, about 4 out of 5 of the homozygous children ($Hb^S Hb^S$) die before they reach reproductive age. Because of this severe disadvantage, one would expect the gene to be eliminated by natural selection and never reach the high frequencies that are observed in some areas. The reason why they do reach them has been explained by Allison. He and others have shown that the heterozygotes are much more resistant to subtertian malaria, the abundance of parasites in the blood being consistently lower and the duration of the disease shorter. Because malaria kills a large number of children, and the heterozygotes are less likely to die from it, they are at a con-siderable advantage over the homozygotes for normal adult haemoglobin $Hb^A Hb^A$. They are also at an advantage over the homozygote $Hb^S Hb^S$ because 4 out of 5 of these die of sickle-cell anaemia. This advantage of the heterozygote results in a stable polymorphism and the allelomorph is found at relatively high frequencies only in areas where there is endemic subtertian malaria. Elsewhere it is eliminated because of the disadvantage of the homozygote. The gene is also, of course, found in some populations which have recently moved from malarial areas, as for example have the American Negroes, amongst whom the gene is quite common.

The situation is slightly complicated in West Africa, for here there is a third allelomorph, Hb^C, producing yet another abnormal red pigment, haemoglobin C. The homozygotes for this gene suffer from anaemia as do the heterozygotes between it and the sickle-cell gene, $Hb^S Hb^C$. Allison's data suggest that in West Africa the heterozygote $Hb^A Hb^C$ is at an advantage over the normal homozygote $Hb^A Hb^A$. This is again probably due to resistance to a blood parasite (not necessarily malarial), although the point has not yet been adequately demonstrated. The data also suggested that $Hb^S Hb^S$ is the most disadvantageous genotype, followed by $Hb^S Hb^C$ and $Hb^C Hb^C$ in that order. Presumably if the other two heterozygotes ($Hb^A Hb^S$ and $Hb^A Hb^C$) are protected from parasites the homozygote for normal haemoglobin $Hb^A Hb^A$ will be the next most disadvantageous. In these circumstances one would expect that where the Hb^S gene is common the Hb^C gene will tend to be rare, and where Hb^C is common Hb^S will tend to be rare. This follows from the fact that the heterozygote $Hb^S Hb^C$ is at a disadvantage because such individuals can suffer from an anaemia similar to sickle-cell anaemia as they have no normal A haemoglobin in their blood. In fact, such a relationship between the frequencies of Hb^S and Hb^C has been found in West Africa. The data therefore support very strongly the view that the polymorphism for these blood pigments is maintained by natural selection in malarial areas because two of the three heterozygotes are at an advantage, one being resistant to malaria, one to malaria or to some other disease, whereas the third heterozygote and two of the homozygotes tend to die from anaemia, and the third homozygote $Hb^A Hb^A$ is more likely to die of malaria and possibly some other infection as well.

It is also interesting to note that in other parts of the world the Hb^S gene is also sometimes found in malarial areas as are other genes producing different abnormal haemoglobins. This suggests that the resistance to malaria is due to the fact that the parasite responsible is not adjusted to utilize these pigments, and therefore does not flourish in individuals in which they are present.

The maintenance of polymorphism by changes in selection

Although many stable polymorphisms are probably maintained by the heterozygote being at an advantage, as in the sickle-cell example and very likely in the case of *Cepaea*, there are other methods by which they can be maintained. For example, it has been pointed out, both by Cain and myself[7] and by Haldane,[35] that in the case of *Cepaea* the polymorphism could be wholly maintained by the uncommon forms being at an advantage because of their rarity, and the common ones being at a disadvantage because of their commonness, although the hypothesis in this instance is unlikely to be true. If thrushes and other birds tended to eat preferentially the commoner conspicuous forms of *Cepaea*, then any very rare variety would be at an advantage because birds would tend not to take it. Consequently the form would increase in frequency until it became common and the birds started to concentrate on eating it. It would then become disadvantageous and proceed to become rarer until it again became advantageous. In such a situation the gene responsible would be at an advantage when rare and at a disadvantage when common, and a stable equilibrium would be reached.

Although this method of maintaining the polymorphism probably plays no important part in *Cepaea*, it is worthy of investigation, for experiments with thrushes show that these birds often exhibit a very marked preference for eating snails of one colour to the exclusion of another if given a choice. Moreover, they will sometimes change their preference and then eat only the colour which they previously rejected. If in the wild their preference was determined by the numbers of the two kinds they saw, and they tended to take the commoner form, a polymorphism could be maintained solely as the result of this feeding habit. This method of maintaining a polymorphism is probably very important in mimicry, discussed in Chapter X, where the frequency of a particular form may determine whether it is recognized as suitable for food by a predator, or whether it is mistaken for an inedible species and not attacked.

Abnormal Mendelian ratios

One interesting and apparently rare method of maintaining a stable polymorphism has been discovered by L. C. Dunn in mice. He found that most laboratory stocks of mice carried, at high frequency, at least one mutant, which, when homozygous, is lethal or renders the individual sterile, but in the heterozygote normally has no detectable effect on the appearance of the mouse. The tail of the animal is, however, absent if one of these mutants is together with one particular allelomorph at this locus. Dunn found a whole series of multiple allelomorphs and showed that different stocks usually possess a different member of the series. The extraordinary thing about the locus is that although Mendelian segregation in the females is normal, so that a back-cross gives a 1 : 1 ratio, in the males it is highly abnormal. With most of the mutants there is a vast excess of the mutant gene amongst the progeny, the heterozygote often producing 95% of individuals like itself. However, at least one of the allelomorphs found in captivity shows, on the contrary, an excess of normals and a deficiency of the mutant in backcross progeny.

In the wild nearly all mouse populations carry at least one of these curious allelomorphs and all the wild ones tested have shown an excess of the mutant form in backcross progeny from the male. This upset in Mendelian segregation means that male heterozygotes will produce an excess of offspring with the mutant amongst their progeny, and, therefore, such an allelomorph will increase in frequency in the population. However, because the homozygote is lethal (or sterile), this increase will be checked when the gene reaches a high enough frequency, for then many lethal homozygotes will be formed, and consequently a stable polymorphism will result. Because the discrepancy from the expected Mendelian ratio can be found experimentally, one can calculate the frequency of the mutant at which stability will be reached. In fact wild populations do not attain stability at the frequency which would be expected from such calculations; this demonstrates that there is selection acting on the heterozygote although its nature has not yet been discovered.

Non-random mating and fertilization

On occasion, mating or fertilization may not be independent of the genetic constitution of the parents. In these circumstances a stable polymorphism can be maintained. In one colony of the Scarlet Tiger Moth (*Panaxia dominula*) there is present, at low frequency (3%), a gene, B^B, which increases the amount of black on the fore and hind wings (*see* Fig. 1, p. 33).[61] Now it has been shown in the laboratory that in this species unlike genotypes tend to mate together rather than with like genotypes.[60] Thus the heterozygote ($B^D B^B$) tends to mate with either homozygote ($B^D B^D$ or $B^B B^B$) rather than with another heterozygote when it has a choice of partners. On the other hand, the homozygotes tend to mate with heterozygotes rather than with homozygotes like themselves.

In this as in many other species of moth the female usually mates only once in her life, but a male can fertilize many females. In the wild the virgin female attracts many males and as many as twenty or thirty can often be seen flying or crawling near her. Consequently if the non-random mating observed in the laboratory also occurs in the wild, the males of the rarer genotype will be at an advantage in competition with others. This follows from the fact that they will usually meet a virgin female of the commoner type and because of the non-random mating will have a better chance of copulating with her than will the other attendant males of her own type. On the other hand, the rarer genotypes will seldom meet a female of their own constitution and so will not often be at a disadvantage in competition with the other males. Consequently the rarer gene will be at an advantage and will increase in frequency until it becomes sufficiently common for this advantage to disappear, at which point a stable situation will ensue. In the case of *dominula* it is doubtful whether the polymorphism is in fact maintained in this way only, for the gene is known to affect other characters such as male fertility, viability of the larvae in the wild, and the likelihood of attack by birds because of its effect on colour.[61] However, if the non-random mating does occur in the wild this will be at least one of the factors tending to ensure stability.

Another quite different example can be taken from plants. The Primrose, *Primula vulgaris* Huds., in common with many other species of the genus, normally has two types of flower (Fig. 7). In one called Pin there is a long style bearing the stigma at the mouth of the corolla tube, half-way down which the anthers are attached (Fig. 7a). Thrum has a short style, so that the stigma is half-way up the tube, and the anthers are at its mouth (Fig. 7b). There are also other morphological and physiological differences, the most important of which, from the point of view of the discussion, is

that any pollen from a Thrum flower, when it arrives on a Thrum
stigma, although it may germinate, does not usually form a pollen
tube, and so does not fertilize the ovules. Pin pollen, on the other
hand, does form a pollen tube, both on a Pin and a Thrum stigma.

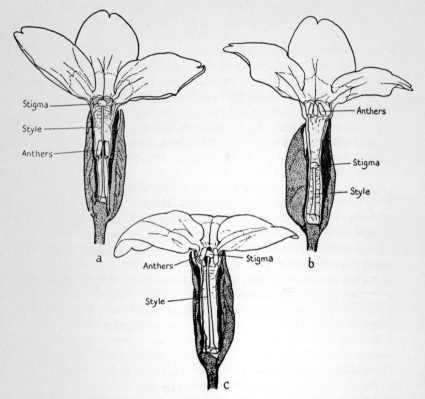

FIG. 7

Sectioned flowers to illustrate heterostyly in the Primrose, *Primula vulgaris*:
(a) a normal Pin flower with a long style and the stigma at the mouth of the
corolla tube half-way down which the anthers are attached; (b) a normal
Thrum flower with a short style and the stigma half-way down the corolla
tube at the mouth of which the anthers are attached; (c) a long homostyle
flower combining some features of both the Pin and the Thrum flower, the
stigma and style, which is long, being like those of a Pin flower but the
anthers are at the mouth of the corolla tube as in the Thrum flower; consequently the stigma is surrounded by the anthers, thus facilitating self-
pollination.

If the flower is Pin, however, the pollen tube grows much more slowly than when it is Thrum, so that if mixed pollen is applied, the Thrum pollen will reach the ovary first and be effective in fertilization. Since all Thrum pollen behaves in the same way, although genetically of two types (Thrum being a heterozygote), it will be realized that the physiological behaviour of the pollen depends not on its own genetic constitution but on that of the plant which produced it. The difference between Pin and Thrum is controlled by a set of closely linked genes which can be regarded for convenience as a single unit (*a super-gene*), since crossing-over between them is very rare. Breeding experiments show that Pin is recessive to Thrum, and because Thrum cannot be fertilized by Thrum all Thrum plants will be heterozygotes. On the other hand, all Pin plants will be homozygotes. Moreover, the physiological and mechanical difficulties of Pin fertilizing Pin, and Thrum fertilizing Thrum, ensure that there will only be cross pollination between the two, as Darwin pointed out (*see* p. 27). Because of the genetic control of the morphological types, each plant will produce about equal numbers of Pin and Thrum progeny, since each mating will be a backcross.

On very rare occasions crossing-over does occur within this complex, and one such cross-over produces what is called a Long Homostyle (Fig. 7c). The corolla and pollen grains appear to be identical with Thrum, and the style is like that of Pin. These plants are self-fertile because the pollen being Thrum-like, and the stigma and style being Pin-like, the physiological situation in self-fertilization is similar to a Pin × Thrum cross. Consequently most of the Homostyle plants will be self-fertilizing, for pollen from the flower will arrive on its own stigma, because of its close proximity to the anthers, even before the flower is properly open. This will almost completely exclude the possibility of Thrum pollen effecting the fertilization of a Homostyle plant. However, the Homostyle pollen will compete on equal terms with the Thrum pollen on the stigma of Pin flowers. The Thrum pollen therefore will be at a great disadvantage to that of the Homostyle, since it can only fertilize Pin flowers whereas Homostyle pollen can fertilize both Pin and Homostyle flowers. This will mean that there will be an increase in the frequency of Homostyles. Moreover, some of these will be heterozygotes and consequently half their pollen will be carrying the Pin gene, so that when the Pin and Thrum plants decrease in frequency, Pin will decrease less rapidly than Thrum.

J. L. Crosby[16] and others have shown that in certain restricted areas of England where Homostyle Primroses are common, the number of Pins is usually much greater than the number of Thrums, as would be expected, whereas in the absence of Homostyles the numbers of Pins and Thrums found are more nearly equal. Crosby gives one count for Somerset of 468 Homostyles, 145 Pins and 15 Thrums.

If no other selective agent were working, one would expect the Homostyle to replace the Pins and Thrums completely and an originally outbreeding population would become inbreeding. However, although the Thrums often disappear when the frequency of Homostyles is high, some Pins are usually present. This is explicable if the Homostyle homozygotes are at a considerable disadvantage to the other three genotypes. To sum up: in the Primrose, populations containing Homostyles reach a stable equilibrium when there are no Thrums but a few Pins present, owing to the competitive advantage of the Homostyles with respect to fertilization and the lowered viability or fertility of the homozygotes. On the other hand, if Homostyles have not been introduced into the population, equilibrium is reached when there are approximately equal numbers of Pins and Thrums owing to the outbreeding mechanism. It seems certain that environmental factors, of which we know nothing, affect the situation, otherwise one would expect all populations to contain Homostyles at high frequency, for, although crossing-over is rare, it must have occurred very many times in the life of the species. It is probable that self-pollination is usually very disadvantageous in this plant but that in exceptional circumstances it may become beneficial, for a time at least, and thus allow Homostyles to establish themselves while the new conditions last.

Self-pollination might be advantageous in an environment where insect pollination is inefficient. This could happen in places where insects are rare, where climatic conditions prevent them from visiting flowers regularly, or where the population of plants is sparse and therefore a single insect seldom visits two different plants of the same species in one day. Moreover, in areas where conditions are severe plant populations are likely to survive only in those few scattered situations in which the environment is particularly suitable. It will usually happen that, at any one time, only a single seed will reach such places not yet colonized by the species. Consequently the resulting plant, if it be self-fertile, will have a better chance of setting seed and establishing a new population than will one that is not self-fertile. It is not surprising therefore that at least one species which has two or more forms of flower over most of its range (is Heterostyled) is Homostyled at the edge of its distribution.

Inbreeding tends to reduce variability and outbreeding to increase it (p. 116) and, since different amounts of variability may be advantageous in different circumstances, a species may change from outbreeding to inbreeding and back again several times in the course of its evolution even when there is no question of one or other breeding system being the more efficient at securing fertilization. Many different mechanisms regulating the breeding system have been evolved in both animals and plants. Besides Heterostyly other methods, such as the production of two sexes (males and females) or an incompatibility system unaccompanied by morphological diversity in flower structure, may be mentioned as mechanisms securing outbreeding. The latter system is found in the Sweet Cherry (*Prunus avium* L.) and the Red Clover (*Trifolium pratense* L.), to mention only two out of a great number of possible examples. The incompatibility results from the fact that there are present in the species a large number of allelomorphs called S_1, S_2, S_3, S_4, etc. Pollen carrying any particular member of this series will not grow sufficiently fast down the style of a plant possessing the same allelomorphs for fertilization to be achieved. Thus S_1 pollen will not fertilize the ovules of a plant of the genotypes S_1S_2, S_1S_3, S_1S_4, etc., but will fertilize individuals that are S_2S_3, S_2S_4 and S_3S_4. Unlike the situation in the primrose the physiological behaviour of the pollen depends on the gene carried by the pollen itself and not on the genotype of the parent plant.

It will be seen not only that self-fertilization cannot take place but also that no homozygotes can be formed. Furthermore in many of the crosses only half the pollen is effective. In a species in which such a system is operating at least three allelomorphs must be present in a population for any seed to be set, since with only two all the genotypes will be the same (S_1S_2) and therefore all the plants will be incompatible with one another. When a particular allelomorph is rare in a population the pollen carrying it will grow properly down most styles but when it is common it will often reach the style of a plant heterozygous for the gene in question. Consequently an incompatibility allelomorph will be at an advantage when rare but at a disadvantage when common, a situation which results in a stable polymorphism. Any new incompatibility allelomorph which arises at the locus will tend initially to become more common so that more and more of them will accumulate in the species and in some more than twenty are known, as for example in the Red Clover.

It will be seen from this brief account that the selection for an outbreeding system often results in a rather marked polymorphism involving many allelomorphs or several groups of closely-linked genes. Such a system can, however, be readily modified by selection, thus allowing a rapid change of breeding system when conditions change.

Gene flow and polymorphism

There are other methods by which a polymorphism can be maintained, but as most of them have not been very fully investigated we need not discuss them here. There is one method, however, which is probably of considerable importance and must be mentioned briefly. If there are two areas in which the habitat is different and one allelomorph is at an advantage in one but at a disadvantage in the other, and there is no migration between the two habitats, the individuals in each area would be expected to be homozygous for the advantageous allelomorph. If, however, there is a limited amount of migration between them, the disadvantageous form in each habitat could be maintained at a modest frequency, and how common it is will be determined by the amount of migration as compared with the selective disadvantage of the gene.

Even if the population covers both habitats, so that there is no separation or sparsely populated area between them, both genes may be maintained. If the organisms are not very mobile or, in the case of the higher plants, the seeds or pollen are not widely distributed by wind, animals or some other means, there will be a gradual change in gene frequency as one moves across the region. The gene which is advantageous in one place will decrease as one moves away from its area of advantage into the region where it is at a disadvantage. Any gradual change in a character from one area to another is called a *cline*, a word suggested originally by J. S. Huxley.[41] A gradual change in the proportion of two or more distinct co-existent varieties from place to place has been called a *morph-ratio cline*.[42]

Such a cline, maintained by natural selection, has been described in the Castor Bean, *Ricinus communis*, by S. C. Harland.[37] There are two different forms, one with a waxy 'bloom' on the stem and the other without it, the latter being recessive. Harland found that the frequency of the two forms at Lima in Peru was determined by the climatic conditions, the form with bloom being at an advantage in warm sunny conditions, and that without it in cooler ones where there is less sun and much fog. The critical factor is the presence or absence of sun with or without fog, and not the temperature. The form without bloom is able to flower and fruit successfully when there are few sunny periods, whereas the other type is not, the inflorescences being formed but then blackening and shrivelling. As a result of this selection and migration by dispersal there is, near Lima, a gradual change in gene frequency from lower to higher elevations so that a morph-ratio cline is present. In the situation studied, fog is present in winter at the lower elevations but conditions are sunny at greater altitudes, so that the frequency of plants with bloom increases from 3 : 1997 at sea level in Lima to 23 : 0 at 7750 feet.

It has been impossible to cover all aspects of stable polymorphism here, or to mention any but a small fraction of the examples of it which have been investigated. It has been necessary to concentrate on a few examples to illustrate some of the ways by which polymorphism can be maintained for a long period of time. The subject has been extensively investigated because the presence of a polymorphism advertises an evolutionary situation repaying close study. Although major genes responsible for polymorphism probably play only a small part in the evolution of a species or higher taxonomic unit (p. 183), the study of them is making an important contribution to our knowledge of natural selection and evolution. There are three main reasons for this:

(i) Polymorphism will only be found where selective agencies are balanced in such a way as to maintain two or more types in a population.

(ii) Because the classes are distinct, they are easy to study and the selective forces operating are likely to be large and therefore easy to detect.

(iii) Even a slight change in selective value will have a marked effect on gene frequencies. In contrast to the situation found when there is polymorphism, an allelomorph at a disadvantage becomes rare and is only maintained in the population as the result of recurrent mutation. When it has become rare even halving its disadvantage (which can about double its frequency; for example from 1 in 10,000 to 1 in 5,000) may initiate a change which is too small to be detected. This circumstance makes the observation of changes in selective values extremely difficult in the absence of polymorphism.

It is worth pointing out that many polymorphic characters have frequently been cited as examples of attributes which have no effect on the survival or reproductive success of their possessors. A knowledge of the principles underlying polymorphism enables one to predict that such characters in fact probably have very large effects on survival and fertility. Even the sickle-cell trait mentioned earlier in this chapter has been given as an example of a neutral or only mildly disadvantageous character. The blood groups in man have also been so described. However, Ford,[28] twenty years ago, pointed out that man was polymorphic for these characters and suggested that they almost certainly had an effect on disease. Nevertheless, most workers denied that they had any special importance except for the capacity of some of them to produce haemolytic disease of the new-born when the foetus is of a different genotype to the mother. It is only since about 1950 that an association between some of the blood groups and common diseases, such as cancer of the stomach, duodenal ulcer, gastric ulcer, pernicious anaemia, and probably diabetes mellitus, has been detected.[14] Had more people been familiar with the principles of polymorphism, some of these facts would have been discovered and investigated years ago.

VI

POLYGENIC INHERITANCE AND
SELECTION

THE last four chapters have been concerned with genes which have large effects and in which segregation can give rise to qualitative variation. Thus in the snail *Cepaea nemoralis* there are a number of distinct forms of the shell. The shells can be yellow, pink or brown, banded or bandless, and if bands are present the number can vary (Fig. 5). In the Peppered Moth (*Biston betularia*) the moth may be black in ground colour or speckled black and white (Fig. 3), and in man the blood group for the so-called ABO system may be A, B, AB or O. However, within these various distinct classes there is often quantitative variation. For example, in the snail the pink shell may be any shade from very pale to deep red, or in the Peppered Moth there may be slight differences in the amount of black present (Fig. 3, c and d). Such quantitative or continuous variation is often the commonest type found in a character. Some or all of it may be caused by differences in the external environment, but very often a part at least is inherited. For example, in man height is affected by the conditions under which the child grows up, but it is also controlled in part by the genotype of the individual.

Genes which individually have small cumulative effects and together control continuous variation are called *polygenes* and have been extensively investigated by Mather. There is no absolute distinction between these and major genes, for genes can have effects intermediate between the easily recognized differences produced by major genes and the minute effects produced by some polygenes. Moreover, those which have one major effect may have others which are very small. For example several genes affecting eye colour in the fruit-fly *Drosophila melanogaster* also have a quantitative effect on the exact shape of the sperma-

101

theca.[58] Nevertheless, despite the absence of a clear distinction between major genes and polygenes, the latter term is useful, and quite adequate for most purposes.

One might think that a continuously varying character such as human height could not be controlled by particulate genes having distinct effects, but H. Nilsson-Ehle showed as early as 1909 that a quantitative character could be controlled in this way if a large number of genes were involved, each by itself having a very small cumulative effect. For example, consider a plant homozygous for only three pairs of genes *AABBCC* affecting seed weight and in which the seeds had an average weight of 54 mg. under particular conditions. (There would be variation around this average weight due to environmental differences.) Now this plant could be crossed to a homozygote *aabbcc* with an average seed weight of 30 mg. If each locus has an equal and cumulative effect and there is no dominance, the heterozygote *AaBbCc* would have an average weight of 42 mg. By crossing two such heterozygotes a number of genotypes will be produced giving various seed weights. The table below gives the number of plants in each of the seven average seed-weight classes which will be obtained if the genes are not linked.

Average seed weight	54	50	46	42	38	34	30
Range found within classes due to environmental factors	48–60	44–56	40–52	36–48	32–44	28–40	24–36
Proportion of plants in each class	1	6	15	20	15	6	1

There will be seven classes, because the substitution of an allelomorph represented by a little letter by one with a big letter gives an increase in weight of 4 mg. However, as the cross gives 27 different genotypes, it is quite obvious that several genotypes must give the same average seed weight. For example *AaBbCc* will give an average of 42 as will also *AABbcc*, *aaBbCC* and several others. This follows from the fact that each possesses three allelomorphs represented by big letters. Therefore each is heavier than an individual carrying none by the same amount, namely 12 mg. Consequently all have an average of 42 mg. and the genotype cannot be ascertained by observing the weight of the seeds. Moreover, there will be variation due to the environment, and the range might well be as great as or even greater than that given in the Table, so that even within one class a certain number of plants will have a seed weight characteristic of another class. The situation becomes even more difficult when there are a large number of genes segregating, and it is not much improved even if there is dominance. Thus even with only three genes the variation becomes nearly continuous and is completely so if the environment causes as great a variation as the substitution of one allelomorph by another.

To put the matter another way, if the variation due to the environment is greater than that due to the individual inherited factors, then no Mendelian ratios will be obtained by breeding. Consequently the situation cannot be investigated by normal Mendelian methods. It is, however, possible to show that the variation is controlled by genes segregating in a Mendelian manner. The transmission of inherited characters by means of genes ensures that the contribution from both parents must be equal (barring certain specific exceptions such as sex-linked genes). Consequently the offspring of reciprocal crosses must be alike if inheritance is Mendelian but not sex-linked. Therefore if the offspring from reciprocal crosses are not alike, factors in the cytoplasm as well as in the nucleus are likely to be involved.

We can go further than just determining the absence of cytoplasmic inheritance and show that nuclear genes are segregating. If the inheritance of a difference is Mendelian, we can predict the consequences of various crosses and test whether our predictions are confirmed. Thus if we take two separate stocks, each of which has been closely inbred for many generations, so that each is homozygous for all loci (*see* p. 116), any variability will be environmental in origin and will not be inherited. On crossing the two lines the resultant F_1 will be heterozygous for all the genes by which the two lines differ, and therefore all the individuals will be genetically alike, but different from their parents. Consequently the variability among them will again be due to the environment.

This situation can be represented from our previous example. Let the two lines be of genotype *AABBCC* and *aabbcc*, in which case the F_1 will be *AaBbCc*. One line (*AABBCC*) will have an average seed weight of 54 mg., but there will be variability due to the environment, in our example giving a range of 12 (48 to 60 mg.) with most of the individuals in the range 52 to 56 and only a few which are heavier or lighter. The other line will have an average of 30 and a range of 12 (24 to 36). The heterozygote will have a mean of 42 and a range of about 12 (36 to 48) with the majority lying between 40 and 44 (*see* Table). These means are different, but the variation or range of weights found is about the same in each. It is true that the range will tend to become greater the more individuals examined, so that in practice we would measure the variation by a statistic called the *variance*. For our present purpose, however, we can use the simpler idea of ranges without invalidating our argument.

In the F_2 generation many new genotypes will be produced, and, as we can see, the combined mean for all of them will be about 42. But now the range of means will be from 30 to 54, and the actual range found, due to environmental effects as well as genetic ones, will be 24 to 60, as can be seen from the Table. Thus we see that the F_2 is very much more variable than the F_1. (This also excludes blending inheritance, for here the F_2 would be less variable than the F_1.) If the next generation (F_3) is obtained, the results will depend on the genotype of the pairs of individuals we choose. The mean weight will vary from family to family and the variability within the families will be greater than in the F_1, but less than in the F_2. If in our example we pick by chance two individuals of genotype *AABBCc* and *AAbbCC*, the offspring will be *AABbCC* with an average weight of 50 and a range from 44 to 56, and *AABbCc* with a mean of 46 and a range from 40 to 52. The total range, from about 40 to 56, is less than in the F_2 but more than in the F_1, or in either of the original parental stocks. We might, however, have chosen two individuals of the constitution *aabbCc* and *aaBbCc*, in which case the weight of the seed of the offspring would have an average of about 36 and a total range between 24 and 48; the range again is less than in the F_2, but the average is different from the other F_3 example. Thus if continuous variation is controlled by the interaction of Mendelian factors (genes) and the environment, inbred lines, on crossing, will give almost the same variability in the parents and F_1 generation, greater variability in the F_2, and in the F_3 greater variability than in the parents or F_1, but less than in the F_2. Moreover, the mean will vary from one F_3 family to another.

The evolution of polygenic systems

The ancestors of most organisms have inhabited a particular type or types of environment for many generations. Consequently they are usually well fitted to their conditions of life and any change in their genetic constitution is likely to be disadvantageous while conditions remain the same. For example, it will be remembered from Chapter I that extremes of weight at birth, whether the infant be too heavy or too light, is disadvantageous (p. 22), the highest survival rate being found among individuals of an intermediate weight of about 8 lb.[43] A similar situation was found by another worker for hatchability in duck eggs, those which were too large or too small showing poor hatchability; that is to say, a departure far from the mean in either direction is disadvantageous.

D. Lack's work[48] brings to light another example of a rather different kind. He investigated the clutch size in a number of species of birds. He found, in some at least, that the mean clutch size was determined by an interaction of genetic and environmental factors. Moreover, he found that, within one species, the larger the clutch size the lower was the survival rate amongst the young. Consequently if the clutch size was small, although survival was high, only a few young were fledged. If the clutch size was very large the number of young that survived was small because there was a large mortality. It was in the class of intermediate clutch size that most young were produced and reared, and, in fact, the best clutch size for giving the maximum number of fully fledged young was close to the average for the species in the area.

In any conditions a fortuitous change in a character cannot usually have better than an even chance of being advantageous, and Fisher[20] has pointed out that the larger the change the more likely it is to be harmful. An organism is a complicated and highly integrated system and therefore a large change in any one part is likely to throw the whole out of balance. Major mutants, because they have large effects, must usually be disadvantageous, and most evolution must be brought about by the accumulation of many small changes. Genetic work on species' and subspecies' crosses tends to support this view, although major genes are sometimes involved in species' and subspecies' differences.[12] Thus we see that much of evolution must depend on the selection of polygenes.

Animals and plants with a short generation time, with more than one generation each year, have a particular problem with regard to being adapted to their environment. This also applies to those with a longer generation time, but which live in a fluctuating environment, so that the conditions under which the parents live are often different from those experienced by their offspring. For example, in *Drosophila* in the spring there will be selection which favours an organism suited to the conditions found at that season. But the offspring from these flies will not be subject to such conditions but will incur those typical of midsummer or autumn. It is clear that in the long run the most successful genotypes will be those adapted to the average of conditions throughout the seasonal cycle.

Because genes can alter each other's effects, an organism, to survive, must have genes which interact appropriately to produce a balanced integrated system, and such an assemblage of genes is called a *gene-complex*. If there is genetic variability and cyclical changes in the environment there will be a tendency for selection to alter the gene-complex away from its optimum during each cycle (because the offspring do not experience the conditions encountered by their parents) so that an absence of such genetic variability is from this point of view desirable. But inherited variability is necessary if there is a permanent change in the environment. These two conditions appear at first sight to be incompatible, but polygenic systems can go a long way towards satisfying both.

Polygenes control continuous variation, so for convenience we can represent an allelomorph at one locus by a $+$ if it increases the character in some direction, or by a $-$ if it decreases it. In a newly established population of cross-breeding organisms, each pair of allelomorphs will be distributed in such a way that the frequency of the three possible genotypes, in the absence of selection, will be $p^2 : 2pq : q^2$, where $2pq$ is the frequency of the heterozygote (Chapter III, p. 57). Moreover, this will happen in one generation. If there are two or more loci which are segregating in the population, not all possible genotypes may be produced in the first generation, even if they are not linked (on the same chromosome). After only a few generations, however, all the allelomorphs will be distributed at random with respect to one another, so that with only two loci and with allelomorphs A and a in the ratio $p : q$ at one, and B and b in the ratio $r : s$ at the other, the frequencies of the genotypes will be as follows: $p^2r^2(AABB)$, $2p^2rs(AABb)$, $p^2s^2(AAbb)$, $2pqr^2(AaBB)$, $4pqrs(AaBb)$, $2pqs^2(Aabb)$ $q^2r^2(aaBB)$, $2q^2rs(aaBb)$ and $q^2s^2(aabb)$. In a new population there will be an increase in variability in each generation until such a stable condition is reached, after which there will be no further change. However, if some of the segregating loci in a population are in the same chromosome it will take very many more generations to reach this equilibrium condition, and the closer the linkage the longer it will take. This follows from the fact that some of the combinations can only be produced by crossing-over between linked loci in an individual heterozygous for both. Let us take our previous example, but now assume the presence of linkage; then if the new population only carried chromosomes $\underline{A\ B}$ and $\underline{a\ b}$, the only genotypes would be

$$\frac{A\ B}{A\ B}\ (AABB),\ \frac{A\ B}{a\ b}\ (AaBb),\ \text{and}\ \frac{a\ b}{a\ b}\ (aabb),\ \text{until a cross-over has}$$

occurred. Such a cross-over, which might not appear for many generations, would give the chromosomes $\underline{A\ b}$ and $\underline{a\ B}$. Consequently although in the end the population will be as variable as the previous one, variability will be released very much more slowly. This slow release of variability is very important, as will be seen later.

It was shown on p. 102 that if polygenes having small additive effects are controlling a character, several different genotypes will produce the same value for this attribute. If the genes are on the same chromosome and we represent different allelomorphs by $+$ or $-$, as suggested above, and not by letters as previously, we can have a series of chromosomes, e.g. (i) $\underline{++-}$ which, in our previous example (p. 102) would have been written *ABc*, or (ii) $\underline{-++}$ (*aBC*), or (iii) $\underline{+-+}$ (*AbC*), giving the following possible genotypes $\dfrac{++-}{++-}$ (*AABBcc*), $\dfrac{++-}{-++}$ (*AaBBCc*), $\dfrac{++-}{+-+}$ (*AABbCc*), $\dfrac{-++}{-++}$ (*aaBBCC*) and so on. From our previous example (Table p. 102), we can say that all will have an average seed weight of 46. Therefore there will be no inherited variability on which selection can operate if only these chromosomes are present. If there had been no linkage there would have been recombination between the loci, and new genotypes with different seed weights would have been produced in the succeeding generation (*see* p. 102). B. J. Harrison and K. Mather[39] have investigated this type of polygenic situation in *Drosophila melanogaster* and shown that a number of individuals which were the same for a particular character had different genotypes with respect to the genes controlling it.

Although the population in our example has no 'actual' inherited variability on which selection could act, it has very considerable 'potential' variability, which, as the result of crossing-over, will be released in time to become actual variability. Thus in an individual of the constitution $\dfrac{++-}{+-+}$ a cross-over can produce the chromosomes $\underline{+++}$ and $\underline{+--}$, or in one which is $\dfrac{-++}{++-}$ the chromosomes $\underline{-+-}$ and $\underline{+++}$ can be obtained. Individuals carrying these new chromosomes can give rise to yet others, and in time it is possible to produce all combinations.

If in our population the best weight is 46, these new chromosomes will tend to be removed from the population by selection, for they will, in most combinations, give seed weights above or below this optimum. In the absence of linkage, selection could not very effectively check the variability because new variability would be released rapidly by recombination. However, if there is linkage between the polygenes, selection can be effective in reducing it because the actual variability is released slowly, and disadvantageous combinations will be removed as quickly as they arise. Moreover, the frequency of recombination is itself under genic control and so can be adjusted by selection (Chapter V, p. 84). Thus linked polygenic systems can evolve and produce a situation in which both the apparently incompatible conditions mentioned on p. 106 are fulfilled. There is little actual variability present at any one time, so that changes in the environment of short duration (a few generations) will produce little effect, and the population can remain fitted to its average environment. There is, however, a vast potential variability which is slowly released and is available for evolutionary change if the environment alters permanently.

As has been pointed out, it is impossible to show the presence of these groups of linked polygenes by normal Mendelian methods, but they can be demonstrated by predicting what effect they will have in breeding experiments, and then demonstrating the presence of the effect. The hypothesis which has been developed suggests that the pressure of natural selection will result in the accumulation of balanced groups of linked polygenes; that is to say, in each chromosome the number of allelomorphs increasing a character and the number decreasing it will be balanced in such a way that the sum total effect will tend to produce the optimum expression of the character in the individual. There is reason to believe that the polygenes affecting a particular character will not necessarily be aggregated in a particular part of a chromosome so that interposed between the loci affecting one character there are likely to be other polygenic loci affecting other characters.

If such a system has been built up it can be broken down by applying prolonged artificial selection. In a population of *Drosophila* with a mean number of about 40 chaetae (bristles), flies can be selected for higher or lower numbers, and Mather has obtained 'high' lines with an average of about 55 and low lines with an average of about 26 from such a stock. If in each generation the two pairs of flies with the highest number of chaetae are bred to give the next generation, progress in increasing the mean number will be rapid at first. However, after only a few generations all the immediately available inherited variability will be lost, because only the chromosomes giving the greatest number of chaetae will be left, the others having been eliminated. After this stage has been reached no further progress will be possible except when new potential genetic variability becomes available as the result of crossing-over within each linkage group. Consequently the mean number will rise sharply each time a new chromosome giving an increased number is formed. However, in the generations following those in which no such crossing-over occurs there will be no progress. If the population has a polygenic system the mean number of chaetae under artificial selection increases by a series of sudden jumps at irregular intervals and not steadily from generation to generation. Mather[51] and his colleagues investigating several characters in *Drosophila* where the number of chromosomes is low have in fact demonstrated such irregular progress under selection, confirming the presence of polygenic systems. In other organisms, for example some plants and animals with many more chromosomes than *Drosophila*, the progress is less irregular because of the larger number of chromosomes. The complete reasons for this cannot be discussed here, but depend on the fact that there will be more polygenes which segregate independently of one another.

It will be remembered that genes controlling different characters are found together in the same chromosome. Therefore, if a particular character is being selected in an experiment not only will new combinations of polygenes affecting this character be selected but also allelomorphs at other loci, affecting different characters, which happen to lie between them on the chromosome. Consequently not only will the character being selected change, but also others not deliberately selected. This has also been observed by Mather and Harrison.[51] These other characters, because they have changed during the course of the experiments, will have departed from their optimum value. This alteration, therefore, will be opposed by natural selection. In many experiments this state of affairs is revealed by a sharp decrease in fertility, so that some lines die out and others can only be maintained by relaxing selection. If the artificial selection is relaxed early in the experiment the character being selected often tends to return to its old value under the influence of natural selection on these other characters. If, however, selection is maintained for long enough, the allelomorphs affecting the characters which are changing but not being selected deliberately will redistribute themselves as the result of crossing-over and selection, so that they will build up new balanced systems. When this has happened, any tendency for the artificially selected character to return to its original value will be opposed by natural selection, for such a change would upset the new balanced systems that have been achieved by the genes affecting the other characters.

Thus we see that integrated polygenic systems result in an 'inertia' which has to be overcome by large or prolonged environmental change before there will be any evolutionary change in the population. Despite the fact that selection is often intense, evolution is usually very slow and two of the reasons for this now become apparent. Firstly, characters are usually well fitted to the environment as a result of the operation of natural selection, so that any change is likely to be deleterious. Secondly, even when there is a change in the environment so that the expression of a character is not at its optimum, any change will tend to be opposed because it will disrupt the balanced system of genes. After a character has changed under the influence of selection, many other characters may also have to be adjusted before any further progress can take place. For example, an increase in height may have to be followed by an increase in the strength of the bone of the leg before any further increase in height will be advantageous. Now a change in each character will necessitate the selection of a new balanced system of polygenes, so that, even when there is quite strong selection for a particular attribute, progress may be slow.

We have been considering how an organism, by building up a balanced polygenic system, can show little actual variability, but still maintain considerable potential variability, which can be utilized if the environment changes permanently. The problem of becoming fitted to different environments can, however, be overcome in other ways. For example, in animals which have a good means of locomotion, such as birds, the organism can choose its own environment so that if conditions change it can move to a more favourable one. Such animals, at least within limits, can survive changing conditions even in the absence of genetic variability. In sessile animals and plants this is not possible, but the same end can be reached if the organism is phenotypically flexible; that is to say, it can change its physiology or its form to fit its environment. Thus many plants which are tall and bushy in sheltered conditions are short in exposed situations, or again many species of plants which are found both in sunny and in shady situations have a different form under the two conditions. Many insects are inconspicuous because they are green in green surroundings and brown in brown ones. This state of affairs is sometimes due to the selection of different gene-complexes in the two situations, but this is by no means always true, for the same genotype can respond differently in different environments. Instances are known in which in the same species a particular variety is sometimes genetically fixed and sometimes an environmental modification, as, for example, the broad leaf form *lobata* of the Cock's Foot, *Dactylis glomerata* L.

Mere variability under different conditions is disadvantageous. It is only advantageous when the organism develops different phenotypes in different environments, and that engendered in each environment is better fitted to it than is any other form produced by a different one. Such a system is clearly particularly suitable when the habitat is very heterogeneous and the organism, unable to choose its environment, must make the best of it.

It is not difficult to see how a gene-complex can be evolved which will give this phenotypic flexibility. A character results from an interaction between the genotype and the environment, so that different conditions tend to produce differences in a character, as for example seed weight or human height, both of which have been discussed previously. Consequently a gene or gene combination which interacts with a particular environment to produce an appropriate effect for that environment will be selected, and genes which do not produce a suitable effect will be eliminated. In this way a stable gene-complex can be built up which produces the correct effect in any environment the organism is likely to encounter, but not necessarily in those which it is not likely to meet, and for which, therefore, there has been no selection. An interesting example which has not yet been adequately investigated is the case of green and brown pupae in Swallowtail butterflies and in Whites of the genus *Pieris*. Not only are two forms of pupae often produced, one green and the other brown, but the green ones which would be likely to be conspicuous in winter tend not to hibernate, the butterflies emerging the same year (p. 67). In other words, a pupa can be the appropriate colour for its own background whether this be green or brown; but, by having a physiological difference, a green pupa will tend to emerge before its background changes to brown in winter, and thus avoid the dangerous situation which would result from a green chrysalis being on a brown background. For then it would be very conspicuous to a predator.

C. H. Waddington[70,71] and his colleagues have investigated the problem of selection for a gene-complex which interacts in a particular way with the environment. They exposed pupae of *Drosophila melanogaster* to a high temperature for a short period of time. Under these conditions a small number of flies appeared which lacked a particular vein on the wing. By selecting these to breed from they increased the proportion of flies showing this character under the treatment in subsequent generations. This is an interesting result, and one that would be expected. What is even more interesting is that after a number of generations some of the flies developed this character even in the absence of heat-treatment, and the proportion of these also increased. In other words, selection had resulted in a character, usually only produced under exceptional environmental conditions, being produced under normal conditions. Selection for those individuals that produced the character only with heat-shock would give a phenotypically flexible stock, whereas selection for those that produced it under both conditions would give a phenotypically fixed one.

This result explains how some plants or animals can develop a gene-complex which produces a particular form fitted to a particular environment under most environmental conditions; that is to say, they are not phenotypically flexible, whereas in others the form is only produced under the appropriate environmental conditions (i.e. it is phenotypically flexible).

When the progeny of an individual usually encounter conditions similar to those experienced by their parents, one would expect selection to produce a gene-complex giving the same characters, suitable for that particular environment, in quite a wide range of conditions, because most variation is likely to be disadvantageous, as pointed out above. However, when the progeny are likely to encounter different conditions from those under which the parents lived, selection will tend to produce a genotype which shows phenotypic flexibility, and develops a character appropriate to its particular habitat. When this is not possible, as in an animal which encounters many in the course of its life, a gene-complex will be produced which is fitted to the average of these environments. Therefore populations of the same species occupying large and diversified areas are usually more alike than populations occupying separate small areas.[19,52]

The effect of the breeding system

So far we have been considering outbreeding species in which individuals are not usually fertilized by a close relative. Many plants and animals are partially or completely self-fertilizing (inbreeding). Now outbreeding promotes heterozygosity but inbreeding leads to homozygosity at most loci (p. 116). Consequently a cross-fertilizing species will evolve a balanced polygenic system adjusted to heterozygosity but an inbreeding species one adjusted to homozygosity. Thus a change from outbreeding to complete inbreeding will usually evolve gradually since it necessitates the acquisition of a new polygenic system. An outbreeding species, if suddenly forced to become inbreeding, will tend to become homozygous. This will disrupt the original polygenic balance and have deleterious effects on viability and fertility. These effects can be sufficiently severe to lead to extinction. Rarely inbreeding may even result in an increase in the frequency of heterozygotes for inversions or translocations. These are likely to be particularly advantageous since, because of the absence of re-combination in them, they will be heterozygous at many more loci than the homozygotes (p. 83), when heterozygosity is at a premium. Little evolution can occur in a complete inbreeder (p. 117), thus explaining why many plants such as the dandelion (*Taraxacum* sp.), although having dispensed with normal fertiliz-ation, have not lost the flower characteristics originally evolved to ensure pollination by insects. Complete inbreeding can be advantageous temporarily but lead to eventual extinction since, unable to evolve rapidly, the species is unlikely to be able to adjust itself to any large and rapid change in the environment.

VII

RECOMBINATION, MUTATION AND GENETIC DRIFT

POPULATION genetics is primarily the study of agents which change the frequencies of genes in populations of organisms. Evolution results from such changes, so that population genetics is fundamental in understanding the causes of evolution. It will have been realized from previous chapters that natural selection is the principal agent in this matter, but there are others which must be considered.

Recombination

Natural selection cannot bring about evolutionary changes if there be no inheritable variability in a population. This was clearly understood by Darwin and demonstrated by W. Johanssen in the early part of the present century. He took mixed stocks of beans which were naturally self-pollinating, and bred from them. These stocks were genetically different from each other, but each was homozygous for all genes as the result of many generations of self-pollination (p. 116). He grew these beans and then in each generation selected for bean size. However, the average in the lines he extracted from his mixture did not change (except in two cases when mutations occurred, *see* below) as the result of his selection despite the fact that there was considerable variability within each line owing to purely environmental factors (*see* p. 102). Had any of this been genetic, then the average size could have been altered by selection. In a mixed stock, a change can be produced, but only up to the point where stocks with the largest or the smallest average seed size have been extracted from the mixture. If Johanssen had artificially cross-pollinated his beans, entirely new combinations of genes would have been

produced in the second generation (p. 104), some of which could
have given plants with an average seed size smaller or larger than
that of the original parents. Consequently if, at the same time,
he had applied selection, it would have been effective as a result of
the increased genetic variability due to the cross-pollination (*see*
Chapter VI). Thus it will be seen that for selection to be effective
there must be genetic variability and that cross-pollination in
plants and cross-fertilization in animals increases this variability
by producing new combinations of genes in each generation.

It will be remembered that Johanssen's stocks were homozygous for all genes as the result of generations of self-pollination.
The reason for this can be easily grasped by performing a simple
experiment using a coin. If we imagine a plant which is segregating for one pair of allelomorphs, *A* and *a*, such a plant will
produce about equal numbers of two types of sex-cells, those
carrying *A* and those carrying *a*. New plants produced by self-
pollination can therefore be either *AA*, *Aa* or *aa*, depending on
which sex-cells come together to form a new individual. We can
see what will happen under these circumstances by representing
A by heads on the coin and *a* by tails, then each time we toss the
coin twice we can find the genotype of an individual in the next
generation. The probability of getting heads on any throw is a
half, so that the chance of getting two running is a quarter. The
chance of getting two tails is the same (a quarter), and one head
and one tail a half. Thus the probability of getting *AA* is a quarter,
Aa a half, and *aa* a quarter. This ratio will be remembered from
Chapter II, and gives the proportions of genotypes in a cross
between two heterozygotes. Consequently approximately half the
offspring from the heterozygotes will become homozygous in
each generation (some being *AA*, some *aa*). A homozygote will
only produce offspring like itself by self-pollination, so that the
proportion of plants still heterozygous will be halved in each
generation. Therefore, after only comparatively few generations,
most lines derived from the original parent will be homozygous.

This will clearly happen at all loci, thus explaining why
Johanssen's beans were not genetically variable but homozygous.
It can be shown that less intense inbreeding (for instance crosses
between cousins) will also reduce the number of heterozygotes
from generation to generation, but much more slowly. Thus

crossbreeding (sometimes called outbreeding) will maintain greater genetic variability in the population than inbreeding, because not only will heterozygotes as well as homozygotes be present, but new combinations will be formed in each generation. There will, therefore, be a larger number of gene combinations found in a crossbreeding population.

Mutation

A change in the average size of the beans in any of Johanssen's inbreeding lines could only have occurred when there was a mutation (a spontaneous change in an inherited factor) of a gene affecting bean size, and in fact this happened twice. Any alteration in size, however, was restricted because after the homozygote for the new allelomorph had been produced there was no more genetic variability present and, therefore, no further advance as the result of selection. Evolution can occur in inbreeding organisms, but it is dependent on the occurrence of new mutations appearing in an appropriate gene-complex, that is, one in which they produce an advantageous effect. Mutations occur very infrequently (p. 52); consequently, evolution will tend to be very slow under an inbreeding system. However, in an outbreeding organism even a few mutations at a limited number of loci will be capable of producing a large number of different gene combinations as the result of segregation and recombination, so that much more rapid and extensive changes can be brought about by selection under a system of outbreeding.

Although mutation provides the genetic variability on which selection can act, it will normally be quite ineffective in bringing about evolution on its own account without the aid of natural selection. Let us again consider only one locus for the time being. If gene A mutates to a, a will increase in frequency only very slowly as mutation is of rare occurrence. If there is also reverse mutation from a to A, a could not replace A in the population, for when it became common A would be produced from it by mutation and a balanced situation would arise, both being present. Moreover, even if there were an extremely small amount of selection against one of the allelomorphs, let us say a, it could never reach a high frequency against this disadvantage. The mutant would increase in frequency until its further increase was exactly counterbalanced by loss due to natural selection. Thus an allelomorph disadvantageous in the heterozygote might arise by mutation at the rate of one in every 25,000 individuals per generation. If there was a population of 2,000,000, we would expect about 80 of these mutants to appear in each generation. Now if 50% of these were lost as the result of selection, less than 80 would be lost if there were not as many as 160 of the mutants in the whole population at the beginning of each generation. If by chance there were more than 160 then more than 80 would be lost. Consequently the mutant would accumulate until there were about 160 of them (reduced to 80 by selection during the life cycle), when an equilibrium would be reached with the loss due to natural selection being exactly counterbalanced by the gain due to mutation.

From what has been said it will be seen that if, in our example, more mutants had been produced in each generation (that is to say, the mutation-rate was higher), the number present when equilibrium is reached will also be higher. Furthermore, if adverse selection is less intense, an equilibrium will again be reached when the mutant is at a higher frequency. Consequently the equilibrium is dependent on the relative values of selection-pressure and mutation-rate. It follows that if the frequency of a mutant is not changing and we know two of the three variables, selection-pressure, mutation-rate and gene-frequency, we can calculate the third. Haldane[33] has used this principle to calculate the mutation-rate for several genes in man. Among others, he has discussed the mutation-rate for a gene which in the heterozygote causes achondroplasia, a condition characterized by the individual being a short-legged type of dwarf. Haldane used data collected by T. Mørch in Denmark. He found that 108 dwarfs had only 27 living children between them, whereas their 457 normal brothers and sisters had 582 children (i.e. the relative proportion surviving $= \frac{27}{108} \times \frac{457}{582} = 0.196$). Now because the dwarfs are heterozygous only half their children will carry the mutants. Consequently the proportion of them surviving compared with their allelomorphs will be $0.196 \times \frac{1}{2}$. It is difficult to estimate the exact frequency of dwarfs at birth, but an estimate from records at the

Lying-in Hospital in Copenhagen gives 10 in 94,075, that is to say 1.06×10^{-4}. It follows from our previous argument that the mutation-rate is $\frac{1}{2}(1-0.098) \times 1.06 \times 10^{-4} = 4.8 \times 10^{-5}$. The factor of $\frac{1}{2}$ appears because each individual carries two allelomorphs at the particular locus, one in each of a pair of chromosomes. The mutant's gene-frequency is therefore $\frac{1}{2}$ that of the dwarfs. Thus if one mutant is found in every 25,000 individuals, one will appear in every 50,000 appropriate chromosomes.

With rare dominant genes almost no homozygotes will be formed. If, however, a gene is completely recessive and only the homozygote is disadvantageous, the mutant will not be at a disadvantage until some homozygotes are formed, so that for a given mutation-rate a recessive gene will reach a higher frequency at equilibrium than a dominant in which the heterozygote has the same selective disadvantage as the recessive homozygote. A mutant which is completely infertile as a heterozygote would be eliminated in each generation, as it could leave no offspring. A recessive mutant, on the other hand, which is sterile only in the homozygote, will not be eliminated in each generation by selection, for the heterozygotes will be fertile. Consequently there will be present in the population not only the surviving mutants produced by mutation in the parents of the population, as is the case with a dominant, but also an additional number which are carried by the descendants of heterozygotes present in earlier generations. It is only when the homozygotes are sufficiently frequent for natural selection to eliminate the same number of mutants in each generation as are produced spontaneously that an equilibrium will be reached.

Genetic drift

There is an agency other than natural selection which can cause a mutant to become common and even to replace its allelomorph completely. The importance of this factor has been advocated by S. Wright,[76] and must be considered here. It will be remembered from Chapter II that in a very large population breeding at random, in which there is no selection or mutation, there will be no change in gene-frequency from generation to generation. In fact, if the gene is not sex-linked and the two allelomorphs are in the ratio of $p : q$, the ratio of the genotypes will be $p^2 : 2pq : q^2$. However, this stability of gene-frequency does not hold if the population is small. As the result of chance alone, it will change from generation to generation, and the smaller the population the greater this change is likely to be. The gene-frequency may fluctuate in such a population until one of the allelomorphs is lost, and then naturally no further change is possible until a new mutant appears.

The way in which, in the absence of selection, gene-frequencies will 'drift' from one value to another can be demonstrated with the aid of a quantity of coloured beads. If we represent one gene A by a red bead and its allelomorph a by a white one, we can represent the gametes (sex-cells) produced by a population by a large bowl full of the beads, with the proportion of red beads representing the proportion of the A gene in the population. If we start with equal numbers of each, we have a gene-frequency for A of 50%. Now all these gametes will not be represented in the next generation, for some will not take part in the process of fertilization, and even of those that do some will not produce offspring which survive to maturity or to breed. We can represent the individuals that reach maturity in the next generation by taking out a handful of these beads. Now it is most unlikely that this handful will have exactly equal numbers of red and white beads, so that the gene-frequency will have changed, and the smaller the handful the greater the departure from a 1 : 1 ratio is likely to be. The individuals in this generation will produce A gametes if they are homozygous for A, a gametes if they are homozygous for a, and equal numbers of A and a gametes if they are heterozygotes. Consequently the genes in the gametes produced by the individuals in this population will be in almost exactly the same proportion as in the population itself. There may be a slight discrepancy but it will be small because very large numbers of sex-cells are produced. We can, therefore, represent the gametes from the population by filling the bowl with red and white beads in the same proportion as they were in our handful. The gene-frequency in the next generation can again be obtained by taking a handful of beads from this new bowl and the process can be repeated. It will be found that the gene-frequency will fluctuate, and if the handfuls are small enough one or other of the allelomorphs will finally be lost. In a population of organisms this process, of course, will apply to all genes for which there are two or more allelomorphs present, if they are not subject to selection.

The experiment can be extended by introducing selection. There are several ways in which this might be done, but perhaps the easiest is to remove a certain proportion of one or other of the allelomorphs (coloured beads) in each generation. If selection is to be on particular genotypes the beads can be put in pairs and then the desired proportion of each type of pair can be removed. There will be three types of pairs, white with white, white with red, and red with red, representing the three possible genotypes. If mating is at random in our population, that is to say the choice of a mate is not determined by its genotype with respect to A and a, the pairs must be made up without reference to the colour of the beads.

It will be found that in large populations (that is, when a very large handful or handfuls of beads are taken out) quite a small amount of selection will determine how gene-frequencies change and what the final frequencies are. If the population is small (a small handful of beads is removed for each generation) drift will play a much more important part in determining the frequencies. They may sometimes change against the direction of selection and the gene being selected may be lost. However, if the experiment is repeated often enough, it will be found that the gene favoured by selection will more often replace its disadvantageous allelomorph than be replaced by it. For example, one might find that in a hundred such experiments the population becomes homozygous for the advantageous gene sixty times and for the disadvantageous one forty times.

If, however, in a small population selection is very intense, that is to say, a large proportion of the disadvantageous genes are removed in each generation, the population will nearly always become homozygous for the advantageous allelomorph. Thus it will be seen that there is a close relationship between the intensity of selection, the size of the population, and the effectiveness of the selection. If the population is very large, quite small amounts of selection will be effective in determining the frequencies of the allelomorphs in the population. If the population is small, and there is little selection, then chance will play an important role.

However, if the population is small and selection is intense, then selection again will be the deciding factor. What has been said for one pair of allelomorphs applies, of course, to all the segregating genes in the population.

The operation of the principles discussed here can be seen in the flower garden. When there are two or more inherited colour forms of a plant which is perpetuated by seed, to avoid losing one or more of the forms it is necessary to have a large stock of them. Even then one form is likely to be at an advantage to all the others because it sets more seed, germinates better or survives better. Consequently, to retain the other forms one has to reduce repeatedly the number of the more prolific form by weeding it out. If, however, not many plants are kept, very careful and extensive selection must be exercised in each generation to avoid ending up with only one form in the flower bed, simply because if, by chance alone, there are too few of any one variety, it may fail to reproduce itself and thus become extinct.

Wright has investigated this chance process, or *genetic drift*, as it is called, using mathematical models, and has applied the principles to the problem of evolution. He originally calculated that in a very small population under constant conditions genetic drift would be of overwhelming importance and that consequently the population would be homozygous for most genes, and there would be little genetic variation. Moreover, he believed that deleterious inherited characters would become established in the population, which would, in consequence, die out and not contribute to the evolution of the species in general. In very large populations, on the other hand, selection would be all-important so that again there would be little genetic variability; the population would be well fitted to its environment, but further evolutionary change would have to wait for the occurrence of advantageous mutations. These would be of rare occurrence, so that evolution would be very slow. In populations of intermediate size both drift and selection would be important, and gene-frequencies would fluctuate. Consequently there would be much genetic variability, new advantageous combinations would be formed on occasion, and therefore evolution would be more rapid than in the other two situations.

It will be remembered that when an allelomorph is lost

from a species it can only be replaced by mutation. But if the species is divided up into a number of populations, in some of which one allelomorph is lost and in some another, then if there is a limited degree of migration between them the gene lost from one population can be replaced from another by migration, thus keeping up the genetic variability. Consequently, Wright maintained that the most rapid evolutionary change would occur in a species which is divided up into a large number of populations of various sizes with a certain amount of migration between them.

Originally, although agreeing that selection could be an important factor in evolution, he seemed to believe that genetic drift was almost essential for prolonged evolutionary changes to occur in a species, and that many of the characters which distinguished them are evolved as the result of genetic drift and are neutral or even deleterious with respect to their effect on the survival of the individual. This view is clearly too extreme, and he has considerably modified it in recent years. When Wright first put forward his hypothesis, others, notably Fisher, maintained that genetic drift was of little or no consequence in evolution.

Population studies

The dispute that arose as the result of these opposed views on drift stimulated many workers to investigate the matter in wild populations. It is only by work in the field that one can get the required information. Mathematical speculations are of great importance in demonstrating what can and what cannot happen under different circumstances, and, therefore, what type of data should be collected in the field. In the absence of data taken from wild populations, they cannot show what in fact happens in nature.

Many genetically controlled characters whose frequencies in populations were once thought to be controlled by genetic drift, such as chromosome inversions in populations of *Drosophila* and the colour and banding patterns in *Cepaea nemoralis*, have now been shown to be controlled largely by selection. The distribution of blood groups in man has also been quoted. Here again there is a growing body of evidence that natural selection is important, for several diseases which can be fatal have been shown to be associated with various blood groups. What is now abundantly clear is that quite small changes in a character can have very large advantages or disadvantages to their carrier, a fact which was not fully realized when Wright first put forward his theory of genetic drift. Moreover, environments are not constant, so that an inherited character which is advantageous at one time may be very disadvantageous at another. For example, there are several forms of the Twin Spot Ladybird, *Adalia bipunctata*, which can be divided up into two groups, the red ones with black spots, and the black ones with red spots. These distinctions are inherited and Timoféeff-Ressovsky[67] in 1940 showed that in the region of Berlin the black form was at a great advantage in summer, and therefore increased in frequency, whereas the red one was at an advantage in winter.

Not all changes in selection occur as rapidly, nor are they so drastic as in the Twin Spot Ladybird, and changes may persist for varying periods of time. Because of this, the frequencies of genes in a population can fluctuate even when it is too big for drift to have anything but a negligible effect. It follows that inherited variability will be kept up and large populations will not be genetically invariable as Wright once thought. Moreover, quite often alternative allelomorphs are both retained even in quite small populations because the heterozygote is at an advantage over the homozygotes, thus maintaining genetic variability. Therefore evolution can occur in large, medium and small populations, and genetic drift is not essential for evolution to take place as Wright originally seemed to believe.

Nevertheless, it is in a species which is divided up into many populations that evolution is most likely to be rapid, for different selective-pressures will be operating in each. Consequently they will tend to evolve in different directions and thus genetic variability in the species will be enhanced. A limited degree of migration between populations will ensure that many new combinations of genes are formed in each generation; some of these at least may lead to evolutionary advance, quite apart from any effects drift might or might not have.

Th. Dobzhansky has suggested that it is meaningless to ask the question: which is the more important, drift or selection, in causing evolution under any given set of circumstances? He maintains that there will be an interplay between these forces so that the question is pointless. However, in the two cases the outcome is different. If selection is playing a major role in a particular species, then there will be a directed change of gene-frequency or a stable balance determined by the nature of the environment. When, over a long period of time, drift is the more important, the direction of evolutionary change will not be related to the nature of the environment, and even slightly disadvantageous characters may become characteristic of the population.

The founder principle

Before leaving the subject of genetic drift it is necessary to consider one aspect of it which has been reiterated by E. Mayr[52] in a slightly different form, and is sometimes known as the founder principle. Mayr argues that a species inhabiting a wide range of habitats with considerable migration between the populations will carry genes which fit the individual well to a large number of habitats, even though they may not produce characters that fit it exceedingly well to any one particular environment. This follows from the fact that the offspring of an individual is likely to encounter a different environment from that met with by the parents, so that the species is adjusted to what one may call the average of the conditions an individual is likely to meet (*see* Chapter VI, pp. 114).

The sum total of the allelomorphs at all the loci in a species, taking into account their frequencies, is often called the *gene-pool*. (Thus to describe a gene-pool accurately one has to specify the gene-frequencies as well as which different allelomorphs are present.) In a species of the type described above, genes which are advantageous under most conditions will be maintained at a high frequency, whereas those which only occasionally find themselves in an environment or gene-complex in which they are advantageous will tend to be rare. Now if a small party of animals or plants from a continent succeeds in colonizing an island, it will not have represented in it all the allelomorphs found in the species. The allelomorphs that are present will be at a different frequency than they are in the species as a whole, as will be seen from the results of the bead experiments (p. 120). But the chance of any one allelomorph finding itself in a suitable gene-complex depends on the frequency of the other genes in the population. Therefore because of a change in gene-frequency, the selective values of the genes will not be the same as in the original population from which they came, quite apart from any change in the external environment in the new colony. There will, therefore, be a change in gene-frequency as the result of natural selection, and a new gene-pool will be evolved which may be very different from that in the species as a whole. In other words there will be an evolutionary change in the population, resulting from a change in the number and frequency of the genes in it owing to its small initial size. Over and above this, there are likely to be other evolutionary changes as the result of natural selection, for the new habitat on the island is unlikely to be either the same or as diverse as the conditions met with by the species on a large continental mass. Thus the island population will become adjusted more closely to the environment, for it does not have to be fitted to such a wide range of conditions.

Mayr believes that this hypothesis may well account for the observed fact that species which have managed to establish themselves on isolated islands tend to evolve striking forms often very different from their continental ancestors, and also that there is considerable variation from island to island.

That the number and types of habitat occupied by a population affects its genetic make-up is strongly supported by the recent work of W. H. Dowdeswell and E. B. Ford.[19] They have investigated the number of spots on the underside of the hind wing of the Meadow Brown butterfly, *Maniola jurtina* (L.). The large islands in the Isles of Scilly all have populations which are apparently identical with one another and do not change from year to year. The smaller islands, however, tend to be very different from one another and from the big ones, even though the population-size on many of them is very large, sometimes of the order of 20,000. The spot-distributions of the insects on all the small islands have remained constant except for one population where there was a large and sudden change (the distribution was constant both before and after it) which coincided with the only big ecological (environmental) change observed in the islands. This change concerned a big alteration in vegetation consequent upon the removal of a herd of cattle.

It seems likely that the spots themselves are unimportant to the survival of the individual. It is more likely that they represent the outward and visible signs of the presence of genes having important physiological effects on which selection is acting. Whether this is true or not, on each small island the environment tends to be fairly uniform, as compared with the diversity on the big ones, but to differ from one island to another. Consequently each population is closely adjusted to the conditions on its island, and tends to differ from other populations. The larger islands, however, have more variation in conditions from place to place so that the population of the Meadow Brown is adjusted to the average of these conditions, which tends to be similar for all the large islands. Thus, as might be expected, the populations on the big islands are, for this reason, similar to one another in appearance and are also similar to the average of all the populations on the small islands which, although each is relatively uniform, represent between them a great range of different environments.

Recent work on polymorphism, for example in ladybirds, snails, flies, moths, mammals and primroses, to mention only a few, suggest that genes having easily recognizable effects are usually subject to very strong selection. Moreover, characters affected by polygenes, such as birth weight in man or egg size in ducks, have been shown to be equally influenced by selection. The evidence at the moment suggests therefore that changes large enough to be detected easily are unlikely to be controlled by genetic drift, except in populations of very small size, which are unlikely to persist for long periods of time because of their size. That selection is controlling the expression of characters does not mean, however, that the polygenes controlling such characters cannot drift in frequency, for several different combinations can produce the same effect, and it is the effect that is being controlled by selection. However, if these genes interact with one another in producing a particular character, a change in their frequency is much less likely to be neutral with respect to its effect on selection. These arguments, of course, do not mean that genetic drift never causes a change in wild populations, but only that it is probably of rather minor importance in evolution.

This view is supported by the fact that characters such as the number of sternopleural bristles or the number of chaetae in

Drosophila do not usually show much change from generation to generation in laboratory stocks which are usually maintained as small populations. Such changes are, however, possible and can be brought about by artificial selection. However, when this is done the stocks tend to become very infertile, indicating that the change is being opposed by natural selection (Chapter VI, p. 110). This suggests that the same effect could not be produced by genetic drift because of the strong selection opposing it.

The reason that there is still controversy about the role played by drift is not due to ignorance of the process. In fact it is the only evolutionary phenomenon which is fully understood because its effect in different circumstances can be determined by exact mathematical methods. The doubt exists because we do not know enough about the other factors which are acting in evolution such as selective values, population-size, migration-rates, mutation-rates, or how genes interact with one another in development. Only more experimental work, particularly field observation, can give final answers to these problems.

VIII

THE EVOLUTION OF DOMINANCE

UP to now we have discussed dominance as if it were the property of a new mutant gene, relative to the normal type, determined when it is formed by mutation. However, it is clear that dominance must be an attribute of a phenotype and not of the gene itself. This is obvious when one remembers that in many instances of stable polymorphism, with the heterozygote at an advantage to both homozygotes, this heterozygote is indistinguishable in appearance from one of the homozygotes. In other words, there is complete dominance for the visual effect. However, had this dominance extended to all the characters controlled by the gene, the heterozygote would have been identical with the homozygote, and, therefore, could not have been at an advantage to it, but would have had the same selective value. Despite the fact that dominance refers to an attribute of a character and not of a gene, it is often convenient to call an allelomorph dominant or recessive simply for the sake of brevity. This I have done on occasion but it must be realized that it is a shorthand notation indicating that one or more particular effects of the mutant are dominant or recessive and that this is not necessarily true for other characters not under discussion.

There are four main hypotheses which have been put forward to account for the fact that the vast majority of mutants are phenotypically recessive to the normal form found in the wild. None of the hypotheses is entirely satisfactory and a short discussion of the difficulties encountered by each will be given after all have been described in some detail. It seems likely that dominance is not always brought about in the same way, and that each hypothesis gives the correct explanation for its presence in at least some instances; therefore all must be considered here.

1. *Fisher's hypothesis*

It was Fisher[20] who first suggested that mutants will not necessarily be completely recessive on their very first appearance in a species and that dominance, like other genetically controlled characters, is modifiable as the result of selection. He pointed out two important facts, namely that (i) mutants are usually disadvantageous and the majority of those appearing in laboratory cultures of animals and plants are recessive. (ii) Mutation is a recurrent phenomenon and most laboratory mutants have a measurable mutation-rate, often of the order of 10^{-5} or less per generation. Therefore in a wild population of a few million individuals, more than 50 or even 100 examples of a particular mutant allelomorph may appear every generation. Fisher argued that, as each of these is disadvantageous, other genes (called modifiers) at different loci, which have the property of reducing the harmful effects of such a mutant, will increase the number of offspring left by its carrier. This will tend to increase the number of generations the gene is likely to survive in the population before it is finally eliminated by natural selection. Consequently these modifiers will be at an advantage in mutant individuals and will therefore be favoured by selection.

Now it will be remembered from Chapter III that if the frequency of two allelomorphs in a population is $p : q$, where $p + q = 1$, the frequency of the genotypes in a large randomly mating population will be approximately $p^2 : 2pq : q^2$ where $2pq$ is the frequency of the heterozygotes. If numerical values are given to p and q, it will be seen that when q is very small the heterozygote is enormously more common than the homozygote of frequency q^2. A disadvantageous mutant will tend to be eliminated by selection and will only be retained in the population as the result of recurrent mutation; consequently its frequency will be very low. Mutants will first appear as heterozygotes. It follows that the mutant homozygote will almost never be formed, and therefore that modifying genes will be selected only with respect to their effect on the heterozygote. As such modifiers accumulate, the heterozygote will come to look more and more like the advantageous homozygote until it is indistinguishable from it. That is to say, the wild type will become completely dominant. During this process the heterozygote will become less disadvantageous as it comes to resemble the normal homozygote and consequently it will be maintained at a higher frequency in the population (Chapter VII, p. 118). The commoner it becomes, the more individuals there will be for selection to act upon, the more effective selection will become in accumulating modifiers and therefore the greater will be the rate at which the character becomes recessive. Two points follow from this argument:

(a) The less disadvantageous the mutant, the higher its frequency in the population in proportion to its mutation-rate, and therefore the more rapidly it will become recessive. Thus mutants which, when heterozygous, are fatal to their carriers, or nearly so, will have a smaller chance of becoming recessive than will less disadvantageous ones because they will be very rare in the population and each will survive for, at the most, a few generations.

(b) The gene will only reach a sufficiently high frequency for an appreciable number of homozygotes to be formed when it has become almost completely recessive.

Fisher points out in support of this theory that when there are two (or more) different disadvantageous allelomorphs present, both being rare, they will almost never occur together in the same individual. That is to say, if the allelomorphs in the population are A, A^b and A^c, the last two being disadvantageous, the combination A^bA^c will almost never be found. Consequently there will be no selection for modifiers of this combination and their effects will show no dominance with respect to one another. This is indeed what is usually observed when such combinations are produced in the laboratory. In contrast to this, there is usually complete dominance between characters controlled by multiple allelomorphs when each is common and the species is therefore polymorphic.

2. *The effect of environmental changes*

Several people have found that severe changes in the environment at a critical period in development will often produce a non-inherited effect which simulates that produced by a known mutant (p. 113). Such non-inherited changes are called *phenocopies*.

For example, in *Drosophila melanogaster* a violent change in temperature during a certain critical stage in pupal life will upset development and causes the appearance of a variety called cross-veinless in some individuals.[70] The same variety is characteristic of flies homozygous for a mutant *cv*. Of course the environmentally produced crossveinless will not be, but the genetically controlled one will be inherited. Now, since environments are not constant, but show wide variation in temperature, humidity, etc., from hour to hour, there must be strong selection for genes whose effect buffers an organism against the abnormal development which results in the production of phenocopies. It is not beyond the bounds of possibility that such genes are also effective in reducing or suppressing the same or similar abnormal development when it is caused by the presence of a new mutant. This may be particularly true when the mutant is present as a heterozygote and, in consequence, produces a less marked effect. It has therefore been suggested that dominance may sometimes be evolved as the result of selection for a gene-complex facilitating normal development under extreme environmental fluctuations because such a gene-complex also tends to maintain normal development when the agent disturbing development is not the external environment but a rare mutant.

Now Waddington,[70] as a result of selection, has produced stocks of *Drosophila melanogaster* which produce a large percentage of crossveinless flies when the pupae have been exposed to heat. It would be interesting to know if, in these stocks, the expression of a mutant *cv*, known to produce crossveinless flies, is enhanced in the heterozygote and whether, in the stock which has been selected for producing a very low proportion of phenocopies, the effect of *cv* is reduced in the homozygote. If the expression of the gene was found to be modified in this way it would support the hypothesis that dominance may sometimes be evolved as a by-product of selection for genes reducing the effect on an organism of sudden changes in the external environment.

3. *Wright's hypothesis*

Wright rejected Fisher's hypothesis for reasons discussed on p. 135 and put forward one of his own.[75] He maintained that most deleterious mutants result from a loss or partial loss of the chemical activity of the gene. This is an old idea which cannot apply to all mutation, for if it did one would have to postulate that the first living organisms had in their nucleus all the genetic potential of higher forms and that this could only be realized by the inactivation of these genes. This view leads to an evolutionary impasse, for beyond a certain point when all genes have been inactivated no more evolution can take place. However, it is not with all mutation that we are at the moment dealing, but only disadvantageous mutants. These can arise as the result of the loss of genetic activity, and that they sometimes do so is supported by a large amount of experimental evidence.

A *catalyst* is a substance which by its presence increases the rate of a chemical reaction. The catalyst is not permanently changed in the reaction, although it may temporarily combine with the substances being changed. Wright suggested that many genes exert their effect by producing *enzymes*, special substances which act as specific catalysts to chemical reactions in living organisms. There is a considerable body of data to support his view that genes do act through enzymes. If this were so, anything which caused the heterozygote to produce an immediate effect identical with that of one of the homozygotes would also ensure that the two were identical in all their subsequent effects; thus there would be complete dominance.

As the result of a mathematical argument, which cannot be discussed here, Wright comes to the conclusion that if the rate of a chemical reaction depends on the concentration of the substance (called the *substrate*) being changed in the reaction, and on the concentration of the catalyst jointly, doubling the quantity of catalyst will not double the amount of the product produced. Furthermore, he concludes that the greater the activity of the gene in producing an enzyme, the more closely will the heterozygote for a less active allelomorph resemble the more active homozygote, because even if the mutant is inactive (the amount of catalyst is halved), the single normal locus in the heterozygote would produce more than half the product present in the normal homozygote. This still applies when there is more than one catalytic process involved in the production of a character. He also points out that the rate of transformation of a substrate often depends also on the concentration of the intermediate compound of substrate and enzyme mentioned above. From this he concludes that if there is an excess of the enzyme, so that all of it is not in a compound with the substrate, an increase in the amount of catalyst will have no effect on the rate of the transformation. Therefore the amount of the product of the chemical process will also not be increased. It follows that a decrease in the amount of enzyme, provided there is still an excess, will also have no effect on the character produced. That is to say, the more active allelomorph will be dominant in effect, provided it and its less active allelomorph together produce an excess of enzyme. Below this critical level of activity there will be an excess of substrate, not of catalyst, and the heterozygote will be somewhat intermediate in appearance between the more and less active homozygotes. Two mutant allelomorphs, both with reduced activity, are therefore likely to show no dominance, one with respect to the other, provided their joint activity falls below this threshold level. Wright therefore concludes that recessive mutants are recessive because they are less active than the normal allelomorph.

4. *Haldane's hypothesis*

Haldane[32] has postulated, as did Wright, that genes produce enzymes and that an excess of enzyme often will produce no greater effect than if just a sufficient quantity of it were present. Thus if a gene produces twice as much enzyme as is required, halving the amount in a heterozygote for a gene producing no enzyme will have no visible effect. This is the conclusion that Wright came to, but Haldane in addition has produced reasons why the wild-type (normal) allelomorph should, in the homozygote, produce twice as much enzyme as is required, a point that Wright's hypothesis does not explain.

He postulates that mutants which produce more enzyme, as well as those which produce less, appear from time to time. These more active genes will not produce a different character and therefore will not be at a disadvantage. However, if the amount of enzyme is reduced by the presence of a less active allelomorph, by environmental influences, or by both, the more active form will be at an advantage as the loss of activity will not result in a change in the character being controlled by the locus. Thus any mutant producing at least twice the minimum amount of enzyme will be selected, and will replace its less active allelomorph, so that there will be a safety factor of 2 or more in the amount of enzyme produced and complete dominance will result.

Objections to the hypotheses

Fisher's hypothesis has been severely criticized on several grounds which must be considered here. Wright emphasizes that the selective advantage of the modifying genes increasing dominance will be very small, being, according to his calculations, of the order of the mutation-rate. Because of the small selective value involved, he objects to the hypothesis on the following grounds:

(i) It will take so long for dominance to be evolved that this can virtually never happen.

(ii) Genes modifying dominance will be likely to have other more important attributes which will determine their frequency in the population, regardless of their effect on the dominance of some other gene.

(iii) When the selective advantage is so small, chance survival (genetic drift) will play such an important role in determining gene-frequencies that selection will be ineffective. This hypothesis is discussed in more detail in Chapter VII.

Haldane agrees with Wright's criticisms and has added a further one of his own.

(iv) He points out that in plants which regularly cross with other members of their own species one would expect, on Fisher's hypothesis, evolution for dominance, whereas in plants which are very rarely cross-pollinated by other plants there would be no tendency, or much less tendency, for dominance to be evolved. This follows from the fact that where mating is not random, and there tends to be inbreeding, the genotypes will not be found in the frequency $p^2 : 2pq : q^2$. There will be, on the contrary, an excess of the homozygotes (Chapter VII, p. 116). Consequently the heterozygote will not be far commoner than the rarer homozygote. Haldane points out that, in fact, in such species dominance seems to be at least as common as in outbreeding forms, which is contrary to expectation on Fisher's hypothesis. It must be remembered, however, that inbreeding species must have evolved from outbreeding ones and the dominance, therefore, may have been evolved in their outbreeding ancestors.

The first three objections are based, in the main, on the belief that the selective values are too small to bring about the necessary evolutionary change. However, the size of the advantage of the modifier is closely related to the frequency of the heterozygotes, so that Wright's and Haldane's criticisms do not apply to situations in which there is polymorphism. Here the heterozygotes will be common and the efficiency of selection, therefore, great. Furthermore, it is now known that many recessive mutants, although usually deleterious, sometimes become common in wild populations, so that the selective values may often be greater than was at first realized. This fact was not known when Fisher first put forward his theory. Sometimes genes, although recessive for their more obvious effects, are not recessive for some of the apparently much more trivial characters which they control. This has been pointed out by Ford, and demonstrated by Schwab[58] for several mutants in the fruit-fly *Drosophila melanogaster*. Ford[23] has argued that these characters are neutral in survival value, and, therefore, have not become either dominant or recessive. On the other hand, the more important disadvantageous characters have become recessive. This interpretation supports Fisher's view.

If we reject Fisher's hypothesis and accept Wright's view that recessive mutants are recessive because they are less active than the normal allelomorph, we have no difficulty over the magnitude of the selective values involved being too small to account for the evolution of dominance. His hypothesis does not, however, really help, for it does not explain why the normal allelomorph is usually so active that even if its activity is halved, it still produces an excess of enzyme. Nor does it explain why such genes are often dominant only for some of their effects, unless one also postulates that the excess of enzyme occurs at some late stage in a chain of chemical reactions, and that the other characters result from the products of earlier steps in the chain. This difficulty also applies with equal force to Haldane's explanation of dominance, but does not affect Fisher's argument where no special mechanism for securing dominance is postulated.

Haldane suggests that the dominance of the wild-type allelomorph results from the selection of allelomorphs forming more enzyme than is normally required. This selection is brought about because, owing to the occasional presence of less active mutants, or of changes in the environment, there is sometimes a reduction in the amount of enzyme produced. In these circumstances, a more active allelomorph will give rise to a normal phenotype, where a less active one will not, even though it may be quite adequate in most circumstances. Ford has pointed out that although the hypothesis seems to differ from that of Fisher, it is not really very distinct, and is only a special case in Fisher's more general hypothesis. He argues that the amount of enzyme or substrate, and therefore the level at which a safety factor of 2 is reached, will be affected by the gene-complex as a whole, so that modifiers can produce the same effect.

Experimental evidence

That dominance can be evolved has been demonstrated on several occasions by selection experiments. Ford[25] obtained specimens of the Currant Moth, *Abraxas grossulariata* (L.), carrying a gene which, when homozygous, changes the white ground colour of the wings to yellow. The heterozygote is intermediate in appearance. He divided his stock into two and, in one, selected from generation to generation for the palest heterozygotes and, in the other, for the yellowest ones. In a few generations the effect of the gene had become almost completely recessive in the pale line, and almost completely dominant in the other line, thereby demonstrating not only that dominance could be evolved, but that this could occur in either direction. The experiment also shows that modifiers are available for producing dominance, as have other similar experiments. Stocks used in the laboratory are usually derived from a few individuals, therefore it is evident that in the much larger populations found in the wild, many more suitable modifiers must be available. Following this work (in order to get a comparison from other material on the number of generations required for the selection) Fisher himself with S. B. Holt[22] got an exactly comparable result (in approximately the same number of generations) working with a 'short-tail' gene in mice.

Although it has been shown that dominance can be evolved, this does not prove that it is normally evolved and that disadvantageous mutants usually become recessive. That they usually do so is, however, indicated by the fact that the common allelomorph found in natural populations is nearly always dominant for its major effects. That is to say, in Haldane's terminology, it has a safety factor of 2. This view is further strengthened by Ford's point that it is often only the more obvious characters controlled by a gene which are, in fact, dominant.

In theory, it is possible to test the hypothesis by putting mutants from one species into the gene-complex of another in which they have not occurred, and seeing if dominance vanishes. There are, however, several difficulties. To begin with, it is almost impossible to be certain that the mutant does not occur, or has not in the past in the test stock, for when two forms can be crossed and produce fertile hybrids they are likely to have arisen from a common ancestral stock in the not too distant past, and the mutant may well have occurred in this stock. The method suggested above has been used in cotton, where S. C. Harland and O. M. Atteck[38] have found some characters which had apparently evolved dominance by the accumulation of modifiers, and others which had apparently become dominant by the selection of more active allelomorphs. However, this interpretation of the data is open to doubt as some races of cotton have more than one pair of each type of chromosome (are polyploid, p. 53), the sets having been increased by doubling during the evolution of the species. Consequently a gene may be represented more than twice and have additive effects which confuse the issue. The final answer to the problem of how the recessiveness of deleterious mutants usually arises is only likely to be reached when we know more about how genes exert their effects. What is certain, however, is that the dominance of the wild type can be brought about by any of the methods discussed here, and that it is often evolved as the result of natural selection.

The importance of the controversy over the method by which dominance is obtained is a direct result of the hypothesis that drift is important in evolution. For, if the recessiveness of deleterious mutants is evolved by the accumulation of genes modifying its effect, as Fisher maintained, very small selective values must usually be effective in controlling evolution, and the great importance of drift in evolution would be thereby disproved.

Polymorphism and the evolution of dominance

As has been pointed out (p. 136), Wright's and Haldane's objections to the theory of the evolution of dominance by the accumulation of modifiers do not apply where polymorphism is involved, for here the heterozygote will be abundant as compared with the situation found when a deleterious mutant is maintained solely by recurrent mutation. Since it is found that the wild type is usually dominant, a new mutant which is advantageous and replaces its old 'wild type' allelomorph almost certainly evolves its dominance if the advantageous characters are not dominant from the beginning. The phenomenon is not easy to study because transient polymorphism by its very nature will not persist for a long period of time (p. 68). At any one moment suitable situations for investigation will be rare. However, at the present time we are fortunate in having a large number of instances of transient polymorphism in the Lepidoptera because of the spread of industrialization (Chapter IV). The genes involved are not ideal for studying dominance because the pigment they produce is black, so that even if double the quantity is produced in the homozygotes, the appearance of the insect will still be black. Therefore, the pigmentation effect may sometimes be dominant from the very beginning even when the homozygotes are producing twice as much pigment as heterozygotes.

Luckily, industrial melanics are not by any means always completely black. There are in existence specimens of the Peppered Moth, *Biston betularia*, which are some of the very first melanics to be caught. A proportion of these have many more white markings on them than are found on present-day heterozygotes taken from industrial areas (Fig. 3, c and d). This at least suggests that during the last hundred years the dominance of black has been evolved. If this interpretation is correct, there has been a surprisingly rapid change which is clearly worth investigation. Moreover, the situation can be and is being analysed by Kettlewell. In parts of Scotland, Devon, Cornwall, Wales and Ireland, which are not industrial, the black form has never been seen. By taking

black forms and crossing them and their black progeny to stocks
from these particular areas, it is possible to put the gene back into
a gene-complex similar to the one present in other parts of
England more than one hundred years ago. It will thus be possible
to see if dominance has been evolved on the lines suggested by
Fisher and, if so, to analyse the situation genetically. For if it has,
the heterozygotes produced in this way should be like the first
melanics caught. A similar investigation can be carried out on
another melanic form (*insularia*) of *B. betularia*, which is far less
black and is controlled by a gene at a different locus (Fig. 3b). It
is already known from Kettlewell's work that, from some areas
at least, a proportion of the heterozygotes for this gene, which
usually resemble homozygous *insularia*, cannot be distinguished
from the pale form. It is important to discover whether in industrial areas more of the heterozygotes can be distinguished from it
than in non-industrial areas.

Stable polymorphism

If dominance can be evolved while an advantageous gene is
replacing its allelomorph, how much more readily may it be
evolved when there is a stable polymorphism! In these circumstances the heterozygote may be maintained at a high frequency
in the population for a very long period of time. Evidence from
fossil snails shows that such stable polymorphism can, in fact, last
for thousands, or even tens of thousands, of years. Moreover,
closely related species often have the same polymorphic forms as,
for example, in *Cepaea nemoralis* and *C. hortensis* or in chim-
panzees and man (p. 77). It is very likely that the ancestral stock,
which in each case gave rise to both species, was also polymorphic
for the same genes. In other words, the polymorphism is older
than either species and so may be the dominance relationships of
the forms.

Ford[27] has recently investigated the polymorphic situation in
a moth, the Lesser Yellow Underwing, *Triphaena comes*. (Hb.).
In most of England the forewings are a pale ochreous-brown, but
in Scotland a dark variety is also sometimes common. This is
controlled by a gene which is semi-dominant. That is to say, the
pale form can nearly always be distinguished from the heterozy-

gote but the heterozygote and dark homozygote cannot always be told apart. The darkest forms are homozygotes and the palest of the dark varieties heterozygotes, but in the intermediate range of shades the two are indistinguishable.

The two forms are found both in Orkney and in the Outer Hebrides. The moths from these areas are isolated from one another by the intervening sea. Ford showed that in stocks from each the dark form behaved as a semi-dominant and that the same gene controls the form in both places. However, when he crossed the two stocks and obtained segregation, the semi-dominance disappeared and there were all grades of intermediates between the pale homozygote and the heterozygote. The experiment demonstrated that the semi-dominance in the two areas resulted from the accumulation of different modifiers and that, when the modifying system was broken down by crossing, dominance disappeared. Here, then, is clear evidence of the evolution of dominance by the accumulation of different modifiers in two isolated populations.

Further evidence in favour of the evolution of dominance comes from some work by E. Caspari.[10] He has investigated a pair of allelomorphs in the flour moth *Ephestia kühniella* (Zeller) which produce red or brown testis-colour respectively. Each allelomorph, however, controls other characters, some of which appear advantageous to their carriers and others disadvantageous. In every instance Caspari found that each allelomorph was dominant, or nearly dominant, for its advantageous characters and recessive for the others. This could hardly be accounted for by postulating different degrees of enzyme activity, for on such a hypothesis one would expect one of the two allelomorphs to be dominant for all its characters, good or bad, and not just for the advantageous ones. In this example the heterozygote is at an advantage to both homozygotes as a direct result of the dominance relationships of the characters, for naturally it only produces the advantageous (dominant) characters of both allelomorphs and none of the disadvantageous recessive ones. The work also illustrates the point that one cannot talk about an allelomorph being dominant, but only a character.

Haldane[32] pointed out that where polymorphism is found it is usually the dominant homozygote which is at a disadvantage to the recessive and not *vice versa* as would be expected on Fisher's hypothesis. Moreover, since when more than one locus is concerned they are often linked, he suggested that these different genes originally arose as the result of the duplication of a small piece of chromosome so that the gene came to be represented more than once in this chromosome. The form resulting from the duplicated segment would be dominant on Haldane's hypothesis, because, as the locus would be represented two or more times in the chromosome, it would produce more enzyme than the equivalent unduplicated segment.

This does not seem to be the correct explanation because of the situation found in *Cepaea nemoralis* and other species ; moreover, the presence of linkage can be explained in another way[62] (p. 84). In *C. nemoralis* there are a number of separate but closely linked genes controlling the colour of the shell, the presence or absence of bands on it, and the presence or absence of pigment in the bands. Other genes controlling the number and width of the bands are also found but are apparently not linked to this series (*see* Fig. 5). Now it has long been known that pink shell-colour is dominant to yellow, and Cain and I[8] have recently been able to show that brown is dominant to both of them. If pink is dominant to yellow because it produces more enzyme (having a safety factor of 2 or more), increasing the amount of enzyme still further should have no effect, but still produce a pink shell on both Wright's and Haldane's hypotheses. Thus, if we believe that brown is produced by yet another duplication of the gene, we must explain how it can produce a brown shell and not merely a pink one, as the duplication should only double the quantity of enzyme present. Moreover, the heterozygote between yellow and brown is brown and not pink as we might have been led to expect if the yellow-brown heterozygote produces about the same amount of enzyme as the homozygous pinks. Furthermore, there are two shades of pink controlled by the locus with pale recessive to dark pink and both dominant to yellow. This relationship is also inconsistent with Haldane's hypothesis. Consequently some further explanation not necessarily involving duplication must be sought to explain the dominance relationships in this and some other polymorphic species. Several are possible, but I will suggest here two which have not, as far as I know, been put forward before. They are as yet only hypotheses.

(i) It is well known that the effects of allelomorphs for which a population is polymorphic do not always differ by the presence of more or less of the same substance. This is clearly seen in the blood groups of man, where each allelomorph usually produces a specific blood-group substance which can be detected by special techniques. The heterozygote usually produces both of these different substances although some produce a third substance not present in either homozygote. Yet again, the gene Hb^A is concerned in the production of the ordinary red blood pigment of man (adult haemoglobin), and its allelomorph Hb^S produces a different haemoglobin (sickle). The heterozygote has about equal quantities of both present in his blood (p. 89). Several other genes, including some of those controlling blood groups, have been found in which the heterozygote apparently produces a substance different from that produced by either homozygote. Thus the AB substance in a person of that blood group is not just a mixture of the A and B substances found in A- or B-type individuals.

Stable polymorphism is often maintained by the heterozygote being at an advantage to both homozygotes. As the visible characters are usually completely dominant or recessive, the advantage of the heterozygote is almost certainly not due directly to a visible character but to some advantageous physiological property correlated with it. Now a number of examples are known in which the dominant homozygote is at a disadvantage both because of poor viability and almost certainly because of its appearance. If, in the beginning, there was no dominance, and the visual effect of the homozygote was deleterious, genes would have been selected which tended to make it look like the more advantageous of the other two phenotypes, which would usually be the ancestral homozygous form. Consequently at first such homozygotes will come to look more and more like their intermediate heterozygotes. A simple method of bringing this about would be to reduce the activity of the gene in the homozygote. However, the advantage of the heterozygote depends on the presence of a physiological property not possessed by either homozygote and therefore dependent on the activity of both types of allelomorph in the same individual. The activity of the two allelomorphs in the heterozygote cannot therefore be much reduced, for this would destroy its advantage and would consequently be opposed by selection. The rare homozygote could come to look very much like the heterozygote, but the heterozygote could not approach the ancestral homozygote in appearance, and dominance of the new form would result. On the other hand, if the heterozygote's disadvantageous appearance can be modified without destroying its physiological advantage, the visible effect of the allelomorph, both in the heterozygote and the homozygote, may be eliminated and no visible polymorphism will be apparent.

The test for this hypothesis is to put the gene into a gene-complex in which modifiers have not been selected, when the visually disadvantageous homozygote should produce an effect more extreme than formerly, and the heterozygote should be intermediate in appearance.

(ii) When the phenotypes of the two homozygotes or alternatively of one homozygote and one heterozygote produced by two allelomorphs are advantageous, but not the other phenotype, it is easy to see that there will be selection in favour of dominance. Thus in mimicry (p. 155), if the new mutant heterozygote is a good mimic, its homozygote will also come to be a good mimic as the result of selection for modifiers, and will therefore resemble the heterozygote. On the other hand, if it is the new homozygote which is the better mimic, it is the heterozygote which will be changed to look like it, and again dominance will be evolved as the result of natural selection. In other words, if there are only two optimum phenotypes, and three genotypes, two of these will come to resemble one another. Moreover, it is the new mutant which will usually come to be dominant in these circumstances if it is being selected on its appearance because the heterozygote must be advantageous from the beginning for the gene to spread at all.

Consequently where the appearance of the varieties is of primary importance (as in mimicry, Chapter X), one would expect to find on this hypothesis of the evolution of dominance that the more recently evolved varieties are dominant to the ancestral type in most examples of polymorphism. This conclusion is supported by the available genetic evidence. Moreover, where there are three varieties, the ancestral A, a more recent form B, and another C with A converted to B by a single factor (gene) and B but not A converted to C by a second factor, B should be dominant to A and C should be dominant to B if the new forms have been selected on account of their appearance. This is the state of affairs found in *Papilio polytes* L. A single gene changes *cyrus*, which is non-mimetic and probably ancestral, to the mimetic form *polytes*, and another gene at a different locus converts *polytes* to the mimic *romulus* but does not affect the appearance of *cyrus*. As expected, *polytes* is dominant to *cyrus* and recessive to *romulus*. If *romulus* is advantageous because of its appearance the gene responsible almost certainly established itself after *polytes* had become common.

Recently Clarke and I have obtained direct evidence suggesting that dominance can be evolved in a polymorphic species. In the mimetic butterfly *Papilio dardanus* (p. 163) we found that of 22 heterozygotes investigated only 7 were intermediate in appearance between the two homozygotes. All of these 22 heterozygotes are formed in nature since the allelomorphs concerned are present together in wild populations. In marked contrast to this result as many as 12 heterozygotes out of 14 produced an intermediate phenotype when the allelomorphs concerned are never found together in nature. Thus dominance is complete in most cases where there has been an opportunity for natural selection to cause the accumulation of modifiers producing it. But where there can have been no such selection, owing to the absence of the heterozygotes in wild populations, it is usually absent! These results strongly support the hypothesis that dominance is usually absent in the beginning but is evolved in polymorphic populations if some of the phenotypes are disadvantageous. They also demonstrate that mimicry can be improved by adjustments in the gene-complex (*see* p. 167).

To sum up, dominance can be present when a mutant appears for the first time, as pointed out by Wright. It can also, however, be evolved as the result of natural selection in the way suggested by Fisher, and also probably by Haldane's method. Only further investigations can finally decide how often, and in what circumstances, dominance is evolved as the result of natural selection, when a rare disadvantageous mutant is involved. When there is polymorphism, dominance is very likely to be evolved, if it is not present from the outset, and this is particularly true when the appearance of the organism is affected by the genes maintaining the polymorphism.

IX

PROTECTIVE COLORATION

Cryptic coloration

The importance of coloration to animals in affording them protection from predators has been stressed in previous chapters. For cryptic (concealing) coloration to be effective, the animal concerned must behave appropriately. This usually means that it must rest on a suitable background and often it has also to orientate itself in a particular way in relation to its surroundings. Thus many moths which sit exposed on the bark of trees habitually align themselves so that the dark markings on their wings lie parallel to the dark cracks in the bark.

A large number of animals are inconspicuous because they are countershaded. That is to say, the upper surface of the body is dark, whereas the lower is light, the two usually grading one into the other. This phenomenon must be familiar to all naturalists and fishermen, for it is widespread among fish, and is so common in this group that it can hardly have escaped the notice of even those who have only examined them on the fishmonger's slab. Countershading is effective in camouflage because it reduces the contrast between the shaded and the non-shaded areas of a solid object in bright light.

Countershaded animals have been used by L. de Ruiter[56] to show the importance of both colour and attitude. He experimented with the larvae of a number of species of moths, including those of several of the British Hawk Moths, the Puss Moth, *Cerura vinula* (L.), and the Kentish Glory, *Endromis versicolora* (L.), all of whose caterpillars are light on the dorsal surface and dark on the ventral one. In these species this is brought about in several ways, including a variation in the exact shade of the green colour of the larvae as well as in the distribution of other pigments, particularly dark ones, on the ventral surface. Anyone wishing to see the extent of the difference need only look at the larva of the common Privet Hawk Moth, *Sphinx ligustri* L. At first sight, one would think that having a dark ventral surface would make the animal more conspicuous in nature where the light comes from the sky above. However, in the wild the caterpillar, when at rest, lies along the *underside* of a twig so that it is the dark ventral surface, and not the light dorsal one, which is uppermost. By exposing larvae to attack by birds, de Ruiter showed how effective the countershading was in making the larva less conspicuous and that, if the caterpillar was incorrectly orientated with respect to the light source, it was more likely to be eaten by the birds.

The larvae of a number of moths closely resemble twigs as a result of their shape, colour and characteristic attitude when at rest, and for this reason are called stick-caterpillars. Poulton, many years ago, discovered that their colour is affected by their environment in such a way that they tend to resemble the background on which they are sitting. Now de Ruiter showed for one such species, the Canary-Shouldered Thorn, *Deuteronomos alniaria* (L.), that his birds could not distinguish at sight the larvae from twigs of the tree on which the caterpillars were feeding. However, the same birds could distinguish between them and twigs from other trees of the same species. This demonstrates in a very clear way that slight differences can have enormous survival values, and that selection for a genotype which enables the animal to adjust its colour to harmonize with its surroundings must be intense.

Selection for variability

Many common animals which are cryptic are also extremely variable, no two individuals being quite alike. This is particularly noticeable among many common species of moth. At first sight one might expect that one pattern would, on the average, be better than all others. In this circumstance selection would result in a species being extremely uniform in appearance. However, although this may be true for many rare animals, in common forms there must often be selection for diversity in the colour patterns present. This results from the fact that vertebrate predators, particularly birds, often concentrate on eating a particular prey if it is very abundant.

Now any naturalist will know that if he is looking for a particular species which is difficult to see, he will become better at finding it after he has found a few individuals and, as it were, 'has got his eye in'. Birds also seem to become more efficient at finding a particular animal after they have encountered it a few times. An individual which has a different colour pattern from the majority may have a better chance of escaping detection, provided it is still sufficiently cryptic, for the bird 'concentrating' on one pattern may miss an animal possessing a different one, as the predator will not have his 'eye in' for the rare pattern. Consequently the rare form will tend to increase in frequency and the commoner form decrease until it becomes sufficiently rare to be advantageous again (*see* Chapter V, p. 91). Moreover, any new pattern can be at an advantage because it is rare; thus there will be selection for variability and no one form will be excessively common.

The variability, of course, will be limited by the available patterns which are cryptic in the circumstances likely to be encountered by the species. If the colour forms are controlled by single major genes, this type of selection will lead to polymorphism. If they are controlled by genes having small additive effects, there will be variability with an array of forms intermediate between the extreme patterns. There can, of course, be a combination of both if there are some major genes and some with less distinct effects. This type of selection may well account for some of the extreme variability found in many very common cryptically coloured animals.

Special resemblance

Not all camouflage is of the sort which makes the individual inconspicuous. The young larvae of many of the Swallowtail butterflies are conspicuous but escape attack because they resemble objects which are not usually a source of food. They are black with a white saddle on the back and resemble a bird-dropping. However, they change their colour pattern radically when they become too large for this method of concealment to be effective. Another animal, a small spider, when frightened rolls itself up to expose its abdomen, and in this attitude looks like the empty head-capsule of an ant. As the spider in question is usually found near ants' nests where such capsules are common, this device may be very effective.

Eye-spots

Many cryptically coloured animals have a second line of defence when disturbed or attacked by a predator. For example, some species of moths and butterflies as well as other insects have large markings on the wings which are imitations of a vertebrate eye. These animals normally rest with the 'eyes' concealed but when disturbed they suddenly expose the spots by moving the wings, and often at the same time make characteristic jerky movements with the whole body. The Eyed Hawk Moth, *Smerinthus ocellatus* (L.), will put on such a demonstration for anyone who cares to disturb it rudely when it is at rest.

A. D. Blest,[4] in a well designed piece of work, has shown not only that the sudden exposure of the previously hidden spots causes small birds to flee, and therefore protects the moth to some extent, but also that the more similar the mark is to an eye in appearance the greater is the protection it affords. The conclusion is inescapable, that its sudden appearance initiates a flight response in the predator similar to that which would be produced by the sudden appearance of a dangerous vertebrate. In other words, the eye-spots mimic the eye of a vertebrate, but not, it will be noticed, any particular species. Blest's results not only show that such a pattern has survival value, but also that an imperfect eye would also be advantageous, though less so. Consequently selection will tend to improve the resemblance of the spot to an eye, and in fact such spots have been evolved from different origins but have converged during evolution, so that the end products are very similar.

The small eye-spots on the underside of the wings of some butterflies are not used to frighten predators in the way described above. They are, however, often excellent imitations of an eye and do protect their possessor, although in a different way. The eye of an animal is often conspicuous and indicates the position of the head, a part of the body which is particularly vulnerable to attack by predators. Consequently when a butterfly such as the Grayling,

Eumenis semele (L.), comes to rest, a nearby predator will tend to attack the eye-spot at the apex of the forewing rather than the vulnerable body of the insect. As the butterfly can fly with a large part of its wing missing, it is likely to survive such an attack. Just this sequence of events has been observed in the Small Heath, *Coenonympha pamphilus* (L.), which, on coming to rest, had its eye-spot attacked by a lizard.

The Grayling has two lines of defence. When it settles on the ground it exposes its eye-spot for a moment and if not attacked covers it with its highly cryptic hind wings. The butterfly, which was at first conspicuous, then becomes extraordinarily inconspicuous, for not only does it match its background well but it also often orientates itself so as to minimize the shadow cast by its wings.

That small eye-spots do deflect the direction of attack away from the body can be seen by examining butterflies with such spots, for it will be found that the part of the wing bearing them is more often missing or more often has the mark of a bird's beak on it than other parts of the wing. C. F. M. Swynnerton,[66] by painting eye-spots on the wings of living insects, releasing them, and examining them on subsequent days, claimed that the spots were attacked more often than other parts of the wing.

Some caterpillars, for example those of several of the Swallowtail butterflies, have small eye-spots despite the fact that all parts of the body could be severely damaged by a predator. They do deflect the direction of attack but instead of this being towards an area where little damage can be done, it is towards one which is especially protected although still potentially vulnerable. In some Swallowtail larvae there are eye-spots on the thorax and close at hand there is a contractile process which is forked and usually yellow or red in colour. This is ordinarily retracted into a pouch behind the head but is suddenly extended when the larva is disturbed by a touch or sometimes even by a shadow. It gives off a strong odour and will cause a stinging sensation if applied to the tongue. This device almost certainly affords some protection to the caterpillar from marauding predators. Thus it will be seen that, in this example, the eye-spots serve to direct attention away from the unprotected abdomen and towards an organ whose properties may cause the attack to be broken off.

Flash coloration

Many insects, such as the Large Yellow Underwing, *Triphaena pronuba* (L.), are remarkably cryptic in appearance when at rest, but when they are disturbed there is a flash of colour as they expose bright markings which until then have been concealed. When they fly away in an erratic manner the colour is very conspicuous and gives a flashing effect as the insect moves. At the end of their flight they usually drop suddenly to the ground and on alighting cover the bright area very quickly, move away a few inches and then remain motionless. The sudden change from extreme conspicuousness to extreme inconspicuousness is very confusing to the observer and it is usually difficult to tell just where the insect has come to rest. In fact it is often impossible to judge its position within several yards. Such a device must often allow the prey to escape when chased by a predator. The cuttlefish, *Sepia*, uses a different and more refined method of escape from predators. It has an 'ink sac' which contains ink composed of granules of melanin pigment. When the animal is attacked it ejects a 'smoke cloud' of this substance and immediately after the emission of the cloud it suddenly becomes very pale in colour and swims at right angles to its original direction of flight. The change of colour together with the appearance of a dark cloud can momentarily confuse the predator so that it attacks the cloud, thus allowing the cuttlefish to escape.

Warning coloration

Not all animals are cryptically coloured, and some are not only very conspicuous but move in such a way that their conspicuousness is accentuated. Instances of such colour patterns and behaviour can be found throughout the animal kingdom. The types of natural selection which have resulted in the evolution of such patterns must be many, not the least important of which, particularly in birds, is sexual selection, discussed at length by Darwin and mentioned briefly in Chapter I. However, at the moment we are concerned with another cause of the evolution of distinctive colour patterns. This is selection for an appearance which gives warning to predators that the animal in question is unpalatable, or possesses some unpleasant or dangerous attribute such as a sting.

It is obviously advantageous for an organism which is relatively inedible or dangerous to advertise this fact as widely as possible, so that it is not repeatedly attacked and thereby damaged. Consequently colours and patterns are evolved which are conspicuous and easily remembered. It is not surprising, therefore, that the patterns tend to be simple and the colours frequently include red, yellow, black or white which, it will be noticed, are also used for road signs as they stand out well against natural backgrounds. Animals which are distasteful but easily damaged are not likely to evolve warning coloration because any new mutant increasing conspicuousness would thereby increase the number of attacks by predators until they had learnt that the animal was unpleasant. If, however, it were not easily damaged, or (because the predator was discouraged by an obnoxious odour before touching it) not damaged at all, a gene increasing conspicuousness would be advantageous. As expected on this hypothesis, the majority of warningly coloured animals, particularly insects, are tough, not easily damaged, and frequently have a strong odour, exude unpleasant substances when handled, or have defensive weapons as, for example, wasps. The larvae of the Buff-tip Moth, *Phalera bucephala* (L.), are gregarious when small and solitary later, and they have an unpleasant smell which when they are very small would be effective only when many were present. Such warning coloration combined with unpleasantness can be found throughout the animal kingdom and is met with in every class of vertebrate animals. For example, the American skunks are warningly coloured black and white, and everyone is aware of their habit of ejecting an evil-smelling fluid at their enemies. The Drongo, *Dicrurus adsimilis* (Bechstein), is a probable example of warning coloration among the birds. The Gila Monster, *Heloderma horridum* (Wiegm.), the salamander, *Salamandra maculosa* (Laur.), and the cat-fish *Plotosus anguillaris* (Bloch) are examples from the reptiles, amphibia and fish respectively.[15]

It is, however, among insects that the phenomenon has been most studied. Everyone will be familiar with the black and yellow stripes on the abdomen of a wasp. That birds can learn to avoid wasps, and once having learnt will remember for a long time (certainly for several months), has been convincingly demonstrated by G. Mostler.[53] It has also been shown that the black and orange larvae of the Cinnabar Moth, *Hypocrita jacobaeae* (L.), are very distasteful.[74] Moreover, it must be quite obvious to everyone who has seen these brightly coloured larvae feeding in their thousands on the heads of ragwort in the summer that they must be unpalatable to birds. If this were not so they would have been eaten, for when there is an abundant and easily obtained food supply, birds tend to 'concentrate' on eating the particular food in question while it is available. Because of this tendency, most gregarious larvae must be protected from predation. Many are hairy and can cause an irritation of the tender parts of the human skin if handled. It is no coincidence that the gregarious Small Tortoiseshell larvae, *Aglais urticae* (L.), are black and yellow, and the equally conspicuous Peacock caterpillars, *Nymphalis io* (L.), are black, both typical warning colours. Only very few birds will eat these larvae. One, the cuckoo, *Cuculus canorus* L., observed feeding on the black, yellow and white larvae of the Scarlet Tiger moth, was found to have the remains of over seventy of these caterpillars in its crop when it was shot. Because a few animals specialize in predating some particular 'protected' species, it has been argued that warning coloration cannot be advantageous, and, therefore, cannot be evolved. However, even if conspicuousness increases the incidence of predation by one species, there will still be selection in its favour if it confers an *overall* advantage by reducing damage inflicted by other would-be predators to an extent sufficient to outweigh the increased mortality from the one.

It will be realized that edible and inedible are relative terms, and anyone who has starved for some time will know that things that he or she would not normally eat can taste very good indeed. This concept of relative edibility must always be borne in mind when discussing warning coloration, and is particularly important when mimicry is being considered (Chapter X).

From the discussion it will be realized that there are many ways in which animals can avoid being detected. They can hide in various places or, if exposed, can be like their background in colour and markings. They may even be conspicuous but resemble some inanimate object which is common in their environment. Moreover, a number, if found, have a second line of defence.

They may frighten their enemy by means of eye-spots or deflect its attack to a non-vulnerable area, thus giving themselves a chance to escape. Still others employ flash coloration which makes it difficult for the predator to find them when they come to rest after being disturbed. On the other hand, species which are not suitable as food for most vertebrates or are especially protected by a sting or some other device have no need to go undetected. In fact it is often to their advantage to be conspicuous so that they are not attacked, but recognized for what they are before there is any likelihood of their being damaged. A more detailed discussion of this subject with numerous examples is given in *Adaptive coloration in animals* by H. B. Cott.

X

MIMICRY

W E have already considered protection by cryptic coloration, the sudden exposure of eye-like markings, flash coloration, resemblance to a dead ant or a bird-dropping, and warning coloration. It seems reasonable to suppose that a relatively palatable species could obtain protection from its enemies if it resembled a warningly coloured and inedible species, and was often mistaken by the predator for this protected form. It was H. W. Bates[2] in 1862 who put forward the hypothesis that this *false warning coloration*, as it is called, is advantageous and can be evolved. He used it to explain the close resemblance in colour pattern between very distantly related butterflies found in South America. The point which impressed him was that although some which were thought to be distasteful had conspicuous colour patterns like their relatives, other butterflies from altogether different taxonomic groups had the same patterns although they were not typical of their own particular genera. It was not long before Wallace showed that the same phenomenon could be found among butterflies in South-east Asia and R. Trimen[69] pointed out that it was also true for Africa (*see* Fig. 8).

Batesian mimicry

This resemblance between edible and inedible species has come to be known as *Batesian mimicry*. Since the warning coloration of the *mimic* is false, whereas that of the *model* (the animal mimicked) is real, it can be predicted that:

(1) The model must be relatively inedible or otherwise protected.

(2) It must have a conspicuous colour pattern, usually one typical of its taxonomic group.

155

(3) It must be common, usually very much more common than the mimic. This condition is obviously necessary because should the model be rare, the predator will have little or no chance of learning that it is protected, and will not, therefore, avoid the mimic. It might even come to associate the pattern with edibility.

(4) Both model and mimic must usually be found together in the same area at the same time. If this were not so, again the predator would have little chance of meeting both, and, therefore, of mistaking the mimic for the distasteful form.

(5) The mimic should bear a very close resemblance to the model. Selection will always tend to increase this resemblance as far as the mimic is concerned, because the closer it is, the less chance there will be of the predator distinguishing between model and mimic. Consequently it will copy the particular pattern of the subspecies of the model with which it is co-existing.

(6) The resemblance only extends to visible structures, colour pattern or behaviour. That is to say, it should deceive the artist (or predator hunting by sight) but not the anatomist.

Müllerian mimicry

F. Müller in 1879 pointed out that a resemblance between species could be evolved in a different way. He argued that young birds learn what is edible and what is not (this is true for most vertebrates), therefore a number of warningly coloured individuals will be killed before the young predators begin to avoid them. Now, if the colour patterns of two distasteful species were sufficiently alike, this loss would be shared between them, and the number of individuals of each species killed would be less than if both patterns had to be learnt separately. This hypothesis has been confirmed. It has now been shown[74] that the black and orange markings on the larvae of the Cinnabar moth can confer some protection on the caterpillars when a predator is not familiar with them, provided that it has previously encountered wasps. Furthermore, in the experiments, when the predator had not attempted to eat either form before, no protection was apparent until some larvae had been tasted.

The characteristics of *Müllerian mimicry* are as follows:

(1) All the species are warningly coloured and protected.

(2) All the species can be equally common.

(3) The resemblance between the forms is not necessarily very exact as it is in Batesian mimicry, because neither species relies on deceiving a predator, but only on reminding it of dangerous or distasteful qualities.

(4) The species are rarely polymorphic.

As has been pointed out previously, the edibility of an object is determined in part by the degree of starvation of the predator. Consequently edibility is only a relative term. It follows that it is not always possible to determine with any certainty whether every particular association is Müllerian or Batesian. Even with the help of the criteria mentioned above, the two types cannot always be distinguished since a Batesian mimic of one species could act as a Müllerian mimic with another. Even though the two types are not distinct, it is possible to say that both are found in many groups of the animal kingdom, including the vertebrates, but especially among the spiders and insects.

Amongst vertebrates may be mentioned mimicry in birds, where it is sometimes well developed. The Black Drongo, *Dricrurus adsimilis*, is a savannah-inhabiting species and is often found in company with a black flycatcher, *Melaenornis pammelaina* Stanley, which is probably a mimic of the distasteful and aggressive Drongo.[15] In South-east Asia there are found a number of Friar-birds of the genus *Philemon*, which are gregarious, conspicuous, noisy and aggressive. Orioles, which are taxonomically very distinct from the Friar-birds, often accompany flocks of the latter, and the two are sometimes very difficult to tell apart. The Friar-birds vary in appearance from region to region, but the Orioles always mimic the local form with which they are co-existing, and not the pattern of a Friar-bird inhabiting some other area. The conspicuousness and aggressiveness of the Friar-birds, as well as the parallel geographical variation in the two groups, makes it almost certain that this is an example of Batesian mimicry in birds, and that the Friar-birds are the models.

Mimicry is more common in the tropics than in the temperate regions, but many of the two-winged flies (Diptera) mimic bees and wasps (Hymenoptera) in temperate climates. In Europe the Broad-bordered and Narrow-bordered Bee Hawk Moths, *Hemaris fuciformis* (L.), and *Hemaris tityus* (L.), and several of the Clearwing moths, for example the Hornet Clearwing, *Sesia apiformis* (Cl.), mimic bees and wasps, as their names imply.

In North America there are a large number of mimics. Among the butterflies the female of the Pipe-vine Swallowtail (*Battus philenor* (L.)) is apparently mimicked by the female of the Parsnip Swallowtail (*Papilio polyxenes asterias* F.), the Spice-bush Swallowtail (*P. troilus* L.) and the black form of the female Tiger Swallowtail (*P. glaucus* L.). Moreover, there are two other butterflies, the Red-spotted Purple (*Limenitis arthemis astyanax* Fabr.) and the female of a fritillary called Diana (*Speyeria diana* (Cram.)) which probably also mimic the Pipe-vine Swallowtail. An even more striking example is given by the Viceroy (*Limenitis archippus archippus* Cram.) which in appearance is amazingly like its model the Monarch (*Danaus plexippus* (L.)).

Objections to the theory

The first and most important type of criticism of the theory of mimicry states that birds and other predators do not eat butterflies, or, anyhow, do not kill a sufficient number to allow mimicry to evolve. However, we can see from the case of Industrial Melanism (Chapter IV) how worthless such statements are, unless backed by a very detailed investigation, which so far they never have been. For until it was observed by Kettlewell,[45] most entomologists and ornithologists denied that birds ever took resting moths off tree trunks. Moreover, the argument about predation does not apply to grasshoppers, flies, spiders, beetles and the host of other animals which show mimicry, as they are well known to be eaten in quantity. Even if this argument is not rejected on these grounds alone, the work of E. B. Poulton and G. D. H. Carpenter has shown that butterflies are eaten by birds. Moreover, the mark of a bird's beak can often be seen on the wings of a butterfly, which shows that they sometimes escape after being attacked.

The second objection is almost the exact opposite of the first. The argument is that a mimic of a model which has a predator specializing on eating it is not protected, but merely draws down on itself the unwelcome attentions of that predator, as for example when the model is an ant, eaten by anteaters, and the mimic a grasshopper. This objection has already been dealt with in Chapter IX, p. 153, in the case of warning coloration.

The third method of attack is to claim that the correspondence between mimic and model is due to the fact that they are living in the same environment so that they develop in the same way. This argument cannot be taken seriously, for often the two are not living in the same environment, although they are living in the same habitat. For example, the environment in which a grasshopper develops and that in which an ant does so must be very different, with regard to food, temperature, humidity and practically any other factor you can think of, although they may well develop in the same habitat, for instance a grass meadow. Furthermore, the resemblance between the model and the mimic is only such as to deceive the eye, there being no corresponding similarity in fundamental body-structure. Thus in one instance the mimic (a young grasshopper) does not have the shape of an ant, but only a dark pattern on its body which resembles one. Finally, we know that in many butterflies different forms are genetically controlled, so that several mimetic forms can develop in the same environment from the same brood simply because they are genetically different.[26] Space will not allow more data to be given here. Anyone wishing to obtain a fuller discussion should consult the book *Mimicry* by G. D. H. Carpenter and E. B. Ford.[9]

The evolution of mimicry

The easiest way of understanding the evolution of mimicry is to study Müllerian and Batesian mimicry separately, because, although they are not always distinguishable in the field, selection in the two instances has different consequences. If there are two distasteful species, A and B, present in an area, and their patterns have some resemblance to one another, they may become Müllerian mimics, as explained on p. 156. Let us now consider the details of the process from the beginning. If the two species with different colour patterns have populations of different sizes, which is likely, a greater proportion of the less numerous species will be eaten by predators before they learn that it is distasteful. For example, if it takes ten attacks for a predator to learn not to take species A it will kill one-thousandth of the population if this is

10,000. On the other hand, if it takes the same number of attacks to learn to avoid species B then, if B has a smaller population of say 100, the predator will take one-tenth of the population before it learns not to take them. Consequently, if a rare mutant should arise in the commoner species A, which converted the visual appearance of A to that of the rarer species B, a greater proportion of the new mutant would be destroyed than that of its non-mutant allelomorph. For by being mistaken for B the mutant is liable to suffer the higher degree of predation just shown to act on this species, whereas the non-mutant will be predated at the lower level found in species A. Consequently the gene will not spread and species A will not evolve towards the appearance of B. However, B will evolve towards A if a mutant changing its pattern to that of A arises, because a smaller proportion of such allelo-morphs will be destroyed by predators, as the mutant individuals experience the degree of predation found in A, not B.

Genes having such a marked effect have little chance of surviving in nature, as has already been pointed out (Chapter VI). A gene producing a less marked effect has a much better chance of surviving. Furthermore, a mutant would receive double protection if it slightly increased the likeness of the commoner species A to the rarer B, providing that it was still mistaken for A but was also sometimes mistaken for B. It would thereby gain an advantage over its allelomorph which is never mistaken for B. Moreover, even if the new mutant is not always recognized as A, it can still spread, provided that it is mistaken for both A and B sufficiently often to outweigh this disadvantage. The consequence of these selection-pressures is that, because of the accumulation of small changes, both species will evolve towards one another in appearance, but the less numerous species will evolve the faster. They will converge as the result of the accumulation of a large number of small changes and there will be no tendency to build up a polymorphism. In fact, the whole trend is towards reducing the number of distasteful forms which the predator has to learn.

Batesian mimicry, on the other hand, is based on deception, the mimic being relatively edible but relying for its protection on being mistaken by a predator for an inedible or dangerous species. In order to understand how this mimicry can be evolved, consider the situation in which, as with the previous example, there is a common species A, obnoxious to its predators and warningly coloured. In the same habitat is found another butterfly, species C, which is much rarer, not warningly coloured, and attacked by predators which avoid A. A mutant appearing in C will be selected if it alters C's pattern in such a way that predators sometimes mistake this new form for species A, and so do not attack it. This is true, of course, only under certain circumstances, for example if the gene does not have some other effect which is deleterious. It will also not be advantageous if the new conspicuous pattern causes the butterfly to be seen more often by predators, which it almost certainly will, unless it is also misidentified for species A sufficiently often to outweigh this disadvantage. Such a situation may rarely occur in highly cryptic animals, where even a slight increase in conspicuousness could prove disastrous, as for example in the case of the stick caterpillars mentioned on p. 147. However, in active animals, particularly butterflies, their activity alone makes them so conspicuous that a change in the colour of a marking from say buff to red would not make them much more obvious when on the wing. Nevertheless, such a drastic change could well render them very difficult to tell apart from a model, particularly when they are moving. This may account, in part, for the commonness of mimicry in animals which move by day compared with nocturnal ones, which rest by day.

Under the conditions discussed above, the edible species C will become mimetic. The selection for a resemblance to the model will also extend to shape, behaviour, odour and any other attributes besides colour pattern by which the predator identifies the prey. While C is evolving towards A, there will be a resultant selection acting on A itself. It will be remembered from the argument of Müller that the predator, when learning to avoid a warningly coloured animal, will kill some of them. If the species has a mimic, the predator will sometimes take one, as they are not easily distinguishable. This animal will be palatable, and consequently it will take longer for the predator to learn to avoid the model, as the pattern will sometimes be associated with an edible, and sometimes an inedible, object. Thus selection may favour any mutant arising in the model which enables the predator to distinguish it from the mimic. It will be seen that the selection-pressure on the model will be greater the commoner the mimic, so that, other things being equal, only a species less common than the model can evolve more rapidly than the model, and, therefore, produce good mimicry. It follows that the accuracy of the mimicry will tend to be greater the rarer the mimic.

There is a point here that is not usually emphasized. The relative abundance of the mimic and model can be affected by the distastefulness of the model. An animal sufficiently obnoxious to make predators avoid its colour pattern, even when having encountered only one, could give protection to a mimetic form as common as, or commoner than, the model. However, if it were much less unpalatable, so that it was normally eaten when food was scarce, it would only protect a mimic much less abundant than itself, for the predators might eat many more before learning to avoid it. If the mimic became too abundant, its enemies would never learn that the model was distasteful, and, therefore, the pattern would no longer protect, and being conspicuous would be selected against.

We may now consider the situation in which a distasteful species, A, and an equally common, or even commoner, edible species, D, co-exist. As before, a new mutant which increases the resemblance of D to A may spread. However, because D is common, and a too common mimic, being conspicuous, is at a disadvantage to its more cryptic form, the new mutant will only spread until it becomes just sufficiently numerous for its initial advantage to be converted into a disadvantage. The non-mutant allelomorph will not, therefore, be completely replaced, and a stable polymorphism will ensue (*see* p. 91). The reasons why the mutant will only increase until it reaches a certain frequency are twofold. Firstly, if the mimetic form becomes too common, the predators will be slow to learn that the model is distasteful and, therefore, will kill more mimics (and models) because of their conspicuousness compared with the non-mimetic form. Secondly, if the mimic is not sufficiently good, and is encountered too often, the predator will learn to distinguish it from the model and it will thereby lose its advantage.

The frequency at which a mimic becomes disadvantageous will be influenced by the abundance of other species which are mimicking the same model. This follows from the fact that it is the ratio of the total number of all the mimics to the models which is important. Consequently, if a model is already mimicked by one or two species, the same colour pattern, if it should subsequently arise in a third species, may not be advantageous whereas it would have been had the other two mimetic species not been present. In other words the abundance of a mimic in one species will affect the selective advantage and therefore the frequency of a similar mimic in another species living in the same area.

Such situations must be quite common since there are often two or more species mimicking the same model in one area. For example, on the west coast of Africa there are two forms of the butterfly *Hypolimnas dubius* (Beauv.), the dominant *dubius*, and the recessive *anthedon*, mimicking *Amauris tartarea* (=*psyttalea*) *damoclides* (Stand.) and *Amauris niavius niavius* (L.) respectively. On the east coast of Africa the forms are somewhat altered; *dubius* here becomes f. *mima* (Fig. 8g) mimicking *Amauris albimaculata* (Btl.) (Fig. 8a), and *anthedon* becomes f. *wahlbergi* (Fig. 8i), mimicking *A.n. dominicanus* (Trim.) (Fig. 8c), the eastern subspecies of the model for *anthedon*. The famous mimetic butterfly *Papilio dardanus* Brown mimics several models including three of those copied by *Hypolimnas dubius*, but unlike *H. dubius* the males are non-mimetic and in some areas the females are also non-mimetic.

In Madagascar, the Comoro Islands and Somaliland the males and females (Fig. 8h) of *P. dardanus* have a ground colour of yellow with black marks on it, have tails on the hind wings and are non-mimetic. Over the rest of the range of the species in Africa, the males resemble fairly closely those just described and are non-mimetic. The females are tailless (with rare exceptions) and are usually mimetic. The mimicry is well developed except in certain parts of Kenya, where the models are rare, and where consequently the mimics, being commoner than the models, are not at an advantage. There are, in this butterfly, a large number of mimetic forms, many of them mimicking the Danaine butterflies. Common ones in South and East Africa are f. *cenea* (Fig. 8d) mimicking *Amauris albimaculata* and *A. echeria* (Stoll.), *hippocoonides* (Fig. 8f) mimicking *A. niavius dominicanus* and *trophonius* (Fig. 8e) mimicking *Danaus chrysippus* (L.) (Fig. 8b). It will be noted that two of these models are the same as those

Fig. 8 Mimicry in African butterflies, for explanation *see text* pp. 163–5.

mimicked by *Hypolimnas dubius*. *P. dardanus* f. *hippocoonides* which mimics *A.n. dominicanus* in the east is replaced in the west by the form *hippocoon* which is similar but mimics *A.n. niavius*, thus exactly paralleling the situation in *H. dubius* (Fig. 8). It is only in South-west Abyssinia, in between the mimetic and non-mimetic races, that mimics with tails are occasionally found.

It will be noticed that the differences between the various forms in both *Hypolimnas dubius* and *Papilio dardanus* as well as other species are controlled by single major genes. During the evolution of the mimicry one would expect that as each new mutant becomes established, other genes, modifying its effect to make it a better mimic, will be selected and become established in the population. However, such modifiers, if they adversely affect the appearance of the non-mimetic or the other mimetic forms, as they may do, will have little chance of spreading, so that most of the modifiers selected will affect only one of the mimetic patterns and not two or more. Thus a new gene-complex improving the mimicry will be established. Consequently there will be evolved a polymorphism for one or more mimetic forms, with single genes 'switching' one pattern to another. Although, if we examine such a situation today, it looks as if the genes produce perfect mimicry from the outset, this is not necessarily true. What has happened is that their original effect has been modified as the result of natural selection in just the same way that the degree of dominance of a gene can be modified (Chapter VIII).

Some people have been much impressed by the fact that a single gene at one step produces almost perfect mimicry. They conclude, therefore, that the mimicry arose as the result of a single mutation, and do not believe that there has been a modification of the original pattern controlled by the gene. R. B. Goldschmidt,[31] who takes up this position, clearly realizes the difficulty of believing that such a good copy could arise by chance alone. He therefore postulates that there are only a limited number of developmental pathways open to a butterfly, so that the number of wing-patterns is restricted, and, therefore, the chance of a new pattern being the same as that of another species is good. Ford[26] has shown that this hypothesis does not account for the facts, and it will only be necessary here to call attention to one or two of the objections. The Indian Swallowtail butterfly *Papilio polytes* f.

romulus mimics the Swallowtail *Polydorus hector* (*L.*). Both have red marks on the wings in similar places, so that it can be argued that, because of a restricted number of possible developmental patterns, both produce similar red pigment in similar places, not to mention the other elements of similarity in the pattern. But the red patches, although similar in shape, position and colour, are chemically quite distinct. Consequently they deceive the artist but not the chemist, exactly as one would expect in mimicry (p. 156). How similar developmental pathways are supposed to give rise to chemically dissimilar reds is not clear. Furthermore, Goldschmidt's hypothesis does not account for the evolution of either cryptic or warning coloration, whereas if one accepts the hypothesis put forward in this chapter, mimicry, cryptic coloration and warning coloration all follow from similar selective pressures as the result of predation, but acting under different circumstances.

Luckily it is not necessary to rely on theoretical arguments, lacking experimental evidence, in order to arrive at the truth, as Darwin so frequently had to do when putting forward his theory of natural selection. Goldschmidt's hypothesis leads to the conclusion that, when there are a number of rather similar subspecies of the model, for example *Amauris niavius niavius* and *A.n. dominicanus*, which are mimicked by different forms of a particular species, these latter will be controlled by a series of multiple allelomorphs. Fisher and Ford's view, on the contrary, leads one to suppose that normally there is only one allelomorph involved and that the differences between the patterns of these mimics are due to modifying genes, which accumulate in different geographical areas as the result of natural selection. Which of these two opposing views is correct can be directly tested by making appropriate matings between such forms. Unfortunately this has only been done on a very small scale. C. A. Clarke and I have crossed forms *hippocoonides* and *hippocoon* of *Papilio dardanus* and obtained no clear-cut segregation of the two forms in the F_2 and backcross generations. This suggests that the two forms are not controlled by different allelomorphs at one locus but that the difference is determined by the presence of modifiers. We were also able to cross form *cenea*, which, like *hippocoonides*, is not found in West Africa, with *hippocoon* from the Gold Coast and found that the offspring which were carrying the gene for *cenea* were intermediate in appearance between *cenea* and *hippocoon*. This is quite unlike the result obtained when *cenea* is crossed with *hippocoonides*, for when this is done normal *cenea* are produced and no intermediates are found (*see also* p. 145).

These breeding results strongly support Fisher and Ford's view since as predicted by their hypothesis there is no clear-cut segregation in the F_2 of a cross between *hippocoon* and *hippocoonides*. Moreover, the results with *cenea* suggest that in West Africa no modifiers have been selected which improve the resemblance between *cenea* and its model, owing to the absence of both in this region. On the other hand, in South and East Africa where both are present such modifiers have accumulated as the result of selection and in consequence the mimicry is good (Fig. 8).

Clarke and I have been able to make a rather similar investigation using the non-polymorphic North American Swallowtail *Papilio polyxenes* whose female mimics the female of *Battus philenor* (p. 158). The difference between this mimic, which is mainly black, and its yellow relatives is due to a single allelomorph with black dominant. However, there are also several related species which are black but non-mimetic. The differences between these and *P. polyxenes* are due to a large number of genes affecting the details of the pattern, notably the distribution of red on the underside of the hind wings and blue on their upper side as well as the distribution of the black pigment itself on all wings. Thus the breeding data taken from both a polymorphic and a non-polymorphic species supports Fisher and Ford's view that mimicry is improved during the course of time by the selection of suitable modifiers.

Many mimics, including *Papilio dardanus*, are mimetic only in the female. The problem of why in polymorphic butterflies the polymorphism is so frequently confined to the females is not a special problem of mimicry, for the same phenomenon is just as common in non-mimetic species. In moths polymorphisms are far less frequently confined to one sex, and when they are it is about as likely to be the male as the female. It has been suggested that the reason for this is that in mimetic species the females require more protection than the males, particularly during egg-laying. This explanation, however, is incompetent to account for the facts because the restriction is found also where mimicry is not involved. Ford has put forward what is probably the correct explanation : butterflies react to visual stimuli in courtship behaviour, as well as to scent, the females having the power of refusal, so that males having a novel pattern may well be less successful in stimulating the female to copulate than those with a more orthodox one. In moths, of course, this does not apply because stimulation is chiefly olfactory. Ford's hypothesis could be tested by dyeing males different colours and finding if they are as successful as males dyed to resemble the normal colour pattern.

Finally, before leaving mimetic polymorphism in butterflies, it seems worth pointing out that there are a number of polymorphic forms in *P. dardanus* and other species, which are non-mimetic and yet remain in the population. This suggests the possibility that the mimicry may not always have evolved as the result of a new gene arising and giving a mimetic resemblance to a suitable model (p. 161). The alternative explanation, which may be true in some instances, is that there was in these species a stable polymorphism with two or more non-mimetic forms as, for example, are known in the Clouded Yellow *Colias croceus* (Fourcroy) and the Silver-washed Fritillary *Argynnis paphia* (L.) in Europe. Then, if a suitable model became available, one of the forms might be converted into a mimic by the accumulation of modifiers. Where a model was not available, the form would remain non-mimetic, thus accounting for the persistence of such forms in mimetic species.

Egg mimicry

Before leaving the subject of mimicry, mention must be made of the similarity often found between the eggs of parasitic birds and their host-species. Several widely distinct groups of birds lay their eggs in the nests of others, and the eggs are incubated and the young raised by foster-parents. This 'brood parasitism' is most highly developed amongst the Old-World cuckoos, and H. N. Southern[64] has written a detailed review of the situation in the European cuckoo, *Cuculus canorus*.

The eggs of this species are smaller, in comparison to the size of the bird, than those in its closest non-parasitic relatives, and than in those cuckoos which parasitize larger host-species, such as crows. The egg is, in fact, about the size of those of its normal hosts, which include Meadow Pipits, *Anthus pratensis* (L.), Reed Warblers, *Acrocephalus scirpaceus* (Herm.), and Redstarts, *Phoenicurus phoenicurus* (L.), to mention only a few. Moreover, in some areas the cuckoo lays several very distinct types of egg, each very closely mimicking the egg of one species of host, so that the two are not readily distinguishable. This is just the situation which would be expected to result from the action of natural selection. An egg which was too large or, more important, judging from both the closeness of the mimicry and behaviour studies on other birds,[68] was of the wrong colour pattern, would tend to be removed from the nest by the host. Alternatively the nest might be abandoned altogether. That this is the correct view has been confirmed by observation, the power of discrimination of the host varying from species to species. No doubt this will be increased by natural selection, for those which do not remove or abandon the cuckoo's egg will rear no offspring of their own over that brood period. Moreover, in those species of cuckoo where the young cuckoo does not destroy the rest of the brood, there is likely to be a higher mortality amongst the young birds owing to competition for food. That young birds can die of starvation when they are nestlings has been shown by D. Lack,[48] for he found that, if the brood was over an optimum size, fewer young survived than with slightly smaller broods. It will be readily appreciated that there will be selection for better and better egg mimicry as the host becomes more and more proficient at detecting the mimetic egg.

There is no difficulty in seeing how egg mimicry can evolve in a species which parasitizes one host or a number of closely related ones with similar eggs. How the evolution of distinct forms is brought about, however, in such species as the European cuckoo, where the males mate with several females, and there are many host-species, is more difficult to see. Southern, in his review, has pointed out that the mimicry is best in those areas where the habitat has been least disturbed by man, and where only a few species are parasitized. He suggests that the cuckoos were divided up into races or *gentes*, and that each evolved in a particular area and became adapted to a particular host. If this be true, the occurrence of two or more gentes in the same area today is due to immigration. However, this does not resolve the difficulty of how the mimicry is maintained, because, where two races are found in the same area, males of one would be expected to mate with females of either, so that there would be a breakdown of the mimicry as the result of the segregation and recombination of genes controlling egg pattern. Southern suggests that the explanation lies in the fact that during their evolution the gentes become adapted to particular types of habitat, which they continue to frequent, so reducing the amount of crossing between them. Such a mechanism reducing crossing between forms is called *ecological isolation*. That is to say, they are separated by the fact that they live in different types of habitat within a region, whereas geographically isolated forms live in different regions.

In support of his hypothesis Southern points out that when the habitat has been much disturbed and diversified by the interference of man, the races would no longer be kept apart, and it is just in these regions that poor egg mimicry is found, for example in parts of the British Isles and Western Europe.

Southern's is probably the correct explanation for the maintenance and local breakdown of the mimicry, but there are other possible explanations, one of which at least should be considered. K. Lorenz[50, 68] and others have shown that some species of bird, particularly geese, when hatched in an incubator behave towards the animal (usually man) which they first see after hatching as if it were their mother, thereafter ignoring their own species, at least for a time. It seems not unlikely that the baby female cuckoo learns its host-species in the same way, and it is known that one female usually lays in one type of nest, except for the rare occasions when a nest is destroyed or deserted just before the cuckoo is due to lay. Consequently, if this hypothesis be true, we can take it that, in a relatively uniform environment with few suitable host-species, any particular cuckoo will be likely to have a very long line of female ancestors, all of whom laid their eggs in the nest of the same host-species. In a more diversified habitat, however, the changes from one host to another are likely to be more frequent because of accidents to nests and the fact that more host-species are available.

During the period when only one host is parasitized there will be selection for increased mimicry of the particular host. If, however, the males are likely to mate with females of different gentes, most of the selection will be ineffective in causing evolution because the males will often be members of other gentes, parasitizing other species. But, as will be remembered from Chapter III, one sex has two sex chromosomes, XX, the other (the male in mammals) has XY, the non-pairing region of Y being passed from father to son, and never into the female line. In birds it is the female which has only one X chromosome, and the male is XX. Consequently if the female cuckoo has a Y chromosome, selection for genes increasing mimicry will be effective for any of them which lie on the non-pairing region of this chromosome. This follows from the fact that they can never be inherited from the male, that is, from a different gens. Unfortunately it is not known whether cuckoos have a Y chromosome, and in most birds it is apparently absent, so that the sex mechanism is XX for the male and X for the female. However, even if most birds have no Y chromosome, this is no argument against the hypothesis for the cuckoo, as it is well known that a Y chromosome can be evolved from the other chromosome pairs, and some animals have several Y's when their relatives have only one, or none.[73] The female cuckoo, therefore, should be examined to see if it has such a chromosome, as the absence of one would disprove the hypothesis.

The mimicry might even be controlled by cytoplasmic inheritance (p. 51) rather than by genes in the nucleus, especially as cytoplasmically inherited particles are known in both plants and animals. The control of a mimetic pattern almost wholly by the female parent is not unknown. In the Swallowtail, *Papilio glaucus*, the wing pattern of all or nearly all of the female offspring is the same as that of their female parent, regardless of the ancestry of the male. Whether this result is due to Y linkage (the female is XY), cytoplasmic inheritance, or some other agency, is not yet known. But that this type of inheritance is known in mimicry makes at least plausible the hypothesis that it also occurs in the cuckoo. Let it be clearly realized however, that the suggestions put forward here are purely speculative, and are given to show that there is an alternative to the view that cuckoos' gentes must be ecologically isolated from one another where there is good mimicry.

Pollination and mimicry in orchids

A very unusual form of mimicry is found in some orchids, particularly those of the genus *Ophrys*, of which the Bee and Fly Orchids are representatives. The labellum or lip of the flower of members of this genus often mimics the appearance of the female of some particular species of the Hymenoptera (bees and wasps) in shape, colouring and hairiness. The phenomenon has been most carefully studied in two mediterranean species of orchid, *O. lutea* Cav. and *O. fusca* Link; the mimicry was found to be an effective device for securing cross-pollination between members of any one species, but to minimize hybridization with other species with which they would be fully fertile if cross-pollination occurred.

Only male bees or wasps are attracted to the flowers and each species of orchid only attracts the males of species whose females are mimicked by the labellum, so that usually no two species of orchid are visited by the same species of Hymenoptera, thus minimizing hybridization. The males are attracted both by the scent and the appearance of the flower. On alighting on it a male is stimulated by the hairs on the labellum and attempts to copulate with the flower and, in so doing, dislodges pollen which it carries to another flower, thus securing cross-pollination.

It will be seen that the mimicry extends not only to the colour and shape of the flower but also to its scent and to the tactile stimulation of the hairs on the lip. Moreover, there are other features of interest which must be mentioned. The Fly Orchid, *Ophrys insectifera* L., is visited and pollinated by the males of a small burrowing wasp *Gorytes mystaceus* (L.) and the Late Spider Orchid, *O. fuciflora* Rchb., by a small bee, *Eucera tuberculata* (F.), but the males apparently only visit the flowers early in their season before the females appear in the area and not thereafter. This suggests that the mimicry is not very good and is only effective when males are sexually highly motivated but unable to find females of their own species with which to copulate.

The efficiency of the mimicry will be maintained by natural selection, since any departure from a sufficiently good resemblance will mean that the flower will not be visited by the male insects and will not be pollinated. That this is a correct deduction is supported by what is known of pollination in the Bee Orchid, *Ophrys apifera* (Huds.). In the British Isles this species is not visited regularly by the males of any species of Hymenoptera, and, as might be expected (p. 96), it is nearly always self-pollinated in this country. Selection for exact mimicry is therefore relaxed and the species tends to be much more variable in flower colour than either the Fly or the Late Spider Orchids. In Southern Europe the Bee Orchid is more often insect-pollinated, as is shown by the presence of a few hybrids between it and other species. It would be of extreme interest to know if it is visited by Hymenoptera of only one species in this region and if so whether it is less variable in appearance than in areas where it is always self-pollinated.

The few hybrids that do occur in *Ophrys* probably result from pollination by insects which do not normally visit the flower of these orchids. Such hybrids are likely to be intermediate between the two parent species with respect to colour, scent and hair stimulation and in consequence to be visited and pollinated by none of the species of Hymenoptera which regularly visit their parents. Consequently such hybrids are almost 'sterile' and are not perpetuated in the population. If, however, there is a species of bee or wasp present whose females are mimicked in a slight degree by the hybrids, the males may visit the flowers and there will be cross-pollination between the hybrids but not between them and their parent species.

Owing to genetic segregation and recombination the progeny will show considerable variability (p. 102) and those most resembling the females of the pollinating insect will be visited, whereas the others will tend not to be pollinated and therefore to be eliminated. In this way a new species could be evolved, and G. L. Stebbins and L. Ferlan[65] have suggested that *Ophrys Murbeckii* Soo arose in this way from a hybrid between *O. fusca* and *O. lutea*. Thus we see that as in other examples of mimicry there are a number of forms which are advantageous but the intermediates between them are very disadvantageous. This type of selection leads to polymorphism in most examples of mimicry because there is no marked tendency for similar forms to mate with one another. However, in the orchids the mimicry does affect the breeding system in this way and so the selection tends to produce speciation (p. 181), not polymorphism (p. 60).

XI

ECOLOGICAL GENETICS

I N previous chapters we have considered natural selection without reference to its effects on the factors determining population-size. Although Wallace laid much emphasis on these controlling factors, they have been frequently neglected by later workers. It is only in very recent times that they have again begun to take the place they deserve in population genetics.

Although population sizes may change quite sharply over short periods of time, it is clear that they usually only fluctuate about an average figure which remains fairly constant over long periods of time. Consequently, each pair of organisms must, on the average, leave exactly two descendants, for, if they left even 2·1 offspring, the population size would increase quite rapidly, whereas if the value were 1·9 the population would soon vanish. This exact balance can only be adequately explained by postulating that as the population increases, either mortality increases or fertility decreases, or both. On the other hand, when there is a reduction in population size, mortality decreases or fertility increases, or again both may occur. The factors which act to bring this about are called *density-dependent* and their operation will tend to keep the population at a constant density.

Wallace believed, as we have seen from the 1858 lecture, that food was the most important of such factors in animals. Ecologists who study the interrelationship of animals, plants and the inanimate environment know, however, of many other density-dependent factors.[48] For example, territories in birds may sometimes act in this way. In some birds the acquisition of an area of ground, or territory, defended from rivals, is essential for breeding success. Consequently whatever the population density, only a limited number of birds can reproduce, even if no other density-dependent factors are acting. The number of young raised in any area is, therefore, not necessarily dependent on the size of the population. Obviously, exceptions to this will occur when a density-dependent factor is not acting, as for example when the population is so small that not all suitable territories are occupied.

The attentions of parasites which can kill or reduce fertility can also control population size. If the population is dense the number of parasites can increase, and so reduce the size or fecundity of the host population by attacking most members of it. When the population becomes reduced in size the parasites will have more difficulty in finding suitable hosts, and will therefore in their lifetime be able to lay fewer eggs, thus causing their own density to decrease. This will allow the host population to increase and under some circumstances it may reach a fairly stable density, under others it may fluctuate about a mean. Bacterial or virus diseases can, of course, act in a very similar manner to parasites.

There are many other possible examples of density-dependent factors, but it will suffice to take just one more. If in particularly inclement conditions an animal has to take shelter in order to survive, the number of suitable refuges may act in a density-dependent way. Those individuals that do not succeed in finding an unoccupied one may perish. When conditions are not so severe, less suitable shelters may be perfectly effective, and consequently under these conditions the population can remain larger. Thus it will be seen that the density of the population can be determined by climatic conditions, but through the agency of density-dependent factors (the number of suitable refuges compared with the number of animals to fill them) which acts differently under different circumstances. It is, of course, true that more than one density-dependent factor may act at any one time, and under changing conditions different factors may come into operation at different times. It is interesting to note that although these factors must be acting in both plants and animals, the plant ecologists hardly ever consider them and it is the animal ecologists who have been responsible for developing the concepts that have been briefly outlined here.

We are now in a position to consider the effect of the spread of a gene as the result of natural selection on the density of a population. If a gene does not interact in some way with a density-dependent factor, the population size will not be materially affected. To explain this we can consider a bird population whose size is determined by the number of suitable nesting-holes. Now, any gene which makes the bird less conspicuous, so that fewer of the individuals carrying it are killed by hawks, will be at an advantage and will spread. As a consequence of this fewer birds will be killed by these predators. However, the number of young produced by the population is dependent upon the number of available nesting-holes. The population will, therefore, not show a marked increase in size as the result of this reduction in attacks by predators. In fact, the breeding population will not be increased at all if all holes were occupied at a lower density, and any increase in the general population will be solely due to an extension of the life span of the individual as a result of reduced mortality from hawks.

The spread of an advantageous gene can even reduce the population size. Haldane[34] has given an example of how this can happen. He took as a hypothetical example a moth whose density was controlled by a specific parasite attacking the larvae. In this example, as is usual with parasites, the host is not found by sight but by some other special sense (e.g. smell). Now if the larvae are also eaten by birds, any gene which makes them less conspicuous, and therefore less often eaten, is selected for. If the effect of this gene is sufficient to reduce the number of larvae eaten by one half, one might, at first sight, expect the population size to increase. The decrease, however, in the destruction of the larvae by birds about doubles the number of larvae present. Consequently the density of the parasites will increase which will result in an increase in the destruction of larvae by them (p. 175). This in turn will decrease the number of hosts (the moths) at all stages in their life cycle. In other words, the spread of an advantageous gene has reduced, and not increased, the population of hosts in the area. Haldane then points out that these deductions can explain why well-adapted species may be rarer than poorly adapted ones. The density of a species will depend on density-dependent factors, and a species which is well adapted to avoiding predators may, as a consequence of this, be rare, if it is controlled by parasites. On the other hand, another species similarly adapted may be common if the controlling factor is attack by predators, not by parasites, for here the increased inconspicuousness reduces the effectiveness of the density-dependent factor at any particular density.

We can therefore see that in an area where the population is controlled by such factors, the spread of a gene as the result of natural selection may (i) have no effect on the density of the population if it does not interact with such factors, (ii) may increase the population if it does interact with them, or (iii) may decrease it under certain circumstances. These conclusions show how close Wallace was to the truth when he discussed this matter. Had he laid more emphasis on characters not affecting density-dependent factors, as well as considering those that do, he might one hundred years ago have come to conclusions similar to those of Haldane.

In areas where the density of a species is not controlled in the way we have discussed above, which may be particularly true of populations on the edge of the species' range where conditions are unfavourable, a colony may depend for its existence on recurrent immigration. In these circumstances the selection of a gene whose effect interacts with a non-density-dependent factor may well allow the species to extend its range. For example, Kettlewell has quoted the distribution of two moths, *Xylophasia hepatica* (Hub.), the Clouded Brindle, and *Miana literosa* (Haworth), the Rosy Minor, in support of this argument. These have been absent from Sheffield for a long time, almost certainly because they could not exist under the severe industrial conditions prevailing. Recently they have reappeared in this area, but the individuals are not of the old form, but are industrial melanics. In other words, the species could not reinvade the area successfully until a suitable gene had arisen. This has increased the average number of offspring left by a pair of moths to a value above 2, whereas previously it must have been below 2. In these circumstances the population will increase until density-dependent factors come into operation and regulate its size.

Haldane[36] has recently pointed out that there are even more important consequences of the effects of genes on density-dependent factors. In the centre of the range of a species such factors will be operating with full intensity. Consequently, as Wallace pointed out in the lecture, any attribute which lessens their effect on an individual will be selected for, since fewer of those with the character will be destroyed at the prevailing density than those without it. Therefore, at the centre of a species' range there will be strong selection for those genotypes which are most resistant to the prevailing density-dependent factors. However, at the edge of the range, where the species is scarce owing to unfavourable conditions, and where it only survives as the result of recurrent immigration, density-dependent factors will not be acting, or even if they are they will be different from those elsewhere. Here there will be strong selection for any forms which are more resistant to the harsh conditions. Because there is continual immigration from the central range of the species, where selection is for resistance to density-dependent factors, the peripheral populations may well be unable to develop a gene-complex adapted to their area; and because of the continual introduction of unsuitable genes which owe their presence to being selected for under quite different conditions, the species will be unable to extend its range to new areas. It is only when the exchange of genes between the central and the peripheral populations is much reduced that the latter will be able to become adapted to the environment.

This principle will apply to all organisms. Haldane has given two examples from moths. He writes : 'Several moth species such as *Peridroma saucia* and *Agrotis ypsilon* winter over in southern England. But large numbers occasionally migrate from the south. This must make it hard or impossible for a race adapted to our climate to establish itself. If the English Channel were as broad as the Straits of Mozambique, rare migrants might have given rise to a local race or subspecies.'

Now, polyploidy is a mechanism which will very much reduce, or even stop, the exchange of genes, because, as explained in Chapter III, polyploids are usually infertile in crosses with their diploid ancestors. If a polyploid arises at the boundary of a species' distribution (for a definition of a species *see* comments, p. 181), the new species so formed may be able to adapt itself to the prevailing conditions, because it will no longer be bombarded with genes advantageous in other areas but disadvantageous in the one in question. It is perhaps noteworthy that in those plants which can perpetuate themselves by asexual means, polyploid forms are frequently found at the edge of the species' range, and in other places where the diploid ancestors do not flourish. The argument developed here would also explain why polyploids are frequently found to be more 'hardy' than their diploid relatives. They will have been able to adapt themselves to their local environment unhampered by the introduction of genes from other areas (called *gene flow*) where conditions are different.

This argument should not be taken as meaning that the production of new species by polyploidy is only important in allowing organisms to adapt themselves to new conditions. Polyploidy, like any other inherited change, produces effects on the organism which may be advantageous and therefore perpetuated. Unlike most other genetic 'mutants', however, hybrids between polyploids and other forms are usually sterile. Consequently, if a polyploid can compete with a diploid successfully, it will survive, and may co-exist with, or eliminate, its diploid progenitor. An advantageous gene, on the other hand, will replace its mutant completely, so that both will not usually be found in the same population, except under the special conditions giving rise to stable polymorphism (Chapter V).

As was explained in Chapter III, polyploidy is generally disadvantageous in bisexually reproducing animals, but in plants any advantages, when it has them, can be exploited more easily. Under very unfavourable conditions, some plants dispense with sexual reproduction and perpetuate themselves by the production of rhizomes, corms, bulbs, etc. In these circumstances polyploidy is not disadvantageous from the point of view of the reduction in fertility. Therefore one would expect polyploidy to be particularly common where the environment is severe, quite apart from any advantage it might have from the point of view of reducing gene flow and allowing the organism to adapt itself to the local environment.

Allotetraploids, unlike autotetraploids, are usually quite fertile (p. 56) and one therefore might expect asexual reproduction to be less important, from the point of view of the polyploid becoming established, than in autotetraploids. An allotetraploid, however, is likely to arise within a population of one or the other of its two parent species most usually in the region where they meet. However, unless it is normally self-pollinating or relies extensively on asexual reproduction to perpetuate itself it is not likely to survive for long. This follows from the fact that if, in the main, it is cross-pollinated most of the pollen it receives, while it is still rare, will be from one or other of its parent diploid species. The resulting seeds will produce triploid plants which will be sterile (p. 56). Consequently the allotetraploid will produce very few allotetraploid offspring and will be unlikely to be able to compete successfully with either of its parent species. For their reproductive capacity, unlike that of the allotetraploid, will not be much impaired owing to the fact that, on the average, the amount of pollen they receive from the allotetraploid will be small compared with the amount they receive from their own species.

Should the allotetraploid establish itself and become common in an area not inhabited by either parent species, or become common within a population of one of its parent species as the result of asexual reproduction, it may exclude the diploid ancestors from the area. The parent species will receive more pollen from the tetraploid than the latter will receive from the parent species when the tetraploid is the commoner. Consequently it will be the diploid parent species which will produce more triploids and therefore have the lower reproductive potential. If there is little cross-pollination from the beginning, or if it is subsequently reduced as the result of natural selection, as it may be (p. 189), the species may be able to co-exist.

In view of the effects of polyploidy, it is not surprising that polyploids often live in different areas from their diploid ancestors and that polyploidy is particularly rare in animals and annual plants reproducing sexually, but is common in perennial herbs with an effective means of vegetative reproduction. It is rare in slow-growing shrubs and trees, the reason for which is not yet fully understood.

XII

THE ORIGIN OF SPECIES

DARWIN was more concerned with natural selection as a factor in causing evolutionary change than with the origin of species as such. In fact he does not even define the term species, although he does indicate that crosses between two of them are often sterile, which is not usually true of varieties.

Darwin was wise to avoid laying special emphasis on species (apart from the rather unfortunate title of the greatest of his books), for, even with our present knowledge of evolution, it is an amazingly difficult concept to define. Those who are interested in the matter should read *Animal species and their evolution* by A. J. Cain. The biological species, which is the one that concerns us here, has never been adequately defined. But in general terms it can be described as a population or group of populations which are capable of exchanging genes, one with another, in nature, if they come into contact. Populations belonging to different species, on the other hand, very rarely hybridize successfully, if at all, when they meet in the wild. This statement holds true for the vast majority of animals and for many plants. Even in plants, where hybridization between species is more common than in animals, crossing must be considerably restricted; for, if this were not so, most groups of species would consist of hybrid swarms, making their separation and classification impossible. Subspecies, often called geographical races, as distinct from species, consist of a number of populations whose individuals possess characters distinguishing them from the members of other populations. However, the members of each are able to hybridize when they meet in nature.

It will be obvious that this criterion of a species does not apply to populations consisting of self-fertilizing or asexually reproducing forms which cannot cross. Nevertheless, such groups often

181

consist of individuals that are very similar, and undoubtedly deserve to be recognized as a unit. Such groups, although they are frequently called species, are of a different kind to those composed of sexually reproducing forms.

The importance of the origin of species in sexually reproducing forms lies in the fact that species exchange genetic material rarely in plants and still more rarely in animals. They are, therefore, committed to separate evolutionary paths. Subspecies, on the other hand, although they may diverge and develop into new species, are not yet committed to this course. For if two such forms extend their range and meet, they may exchange so much genetic material that they form one single unit. In this way, any distinctness or specialization they have evolved may be lost (*see also* p. 188).

As will be remembered from Chapter I, Lamarck believed that animals and plants changed gradually as a result of the inheritance of acquired characters. This hypothesis may be true to a limited degree in bacteria, and perhaps even unicellular animals and plants. All the evidence suggests, however, that it is unimportant in the evolution of the higher animals and plants.

Darwin, like Lamarck, maintained that species evolved gradually, but he suggested that this resulted from the action of natural selection. This view is incompatible with the hypothesis of blending inheritance (p. 23), and it is only with particulate inheritance, first demonstrated by Mendel and unknown to Darwin, that natural selection can be effective. Although the discovery of particulate inheritance removed the most serious objection to Darwin's views, early geneticists believed that the new discoveries were fatal to them. They pointed out that new and striking varieties arise suddenly by mutation, and suggested that species are generated in the same way. De Vries noted sudden changes in a species of the Evening Primrose, and concluded that such 'mutations' gave rise to new species at a single stroke. To him species differed by single major genes which would segregate if a fertile cross could be obtained. It has since been shown that the results he got in his work were, in the main, due to chromosome translocations (p. 53) in the species he studied, although he also observed one difference that was due to the production of a polyploid.

The early Mendelians also supported their argument against the effectiveness of natural selection by stating that the discrete changes controlled by genes could not give rise to continuous variation, and that therefore evolution could not proceed by a series of minute changes. This idea was exploded by Nilsson-Ehle, Fisher and others, who showed that genes can control continuous variation.

Fisher also showed that the probability of a new and very distinct species arising at one step is exceedingly remote. Any change in a highly integrated organism is likely to be deleterious, and the larger the change the more likely it is to upset the animal's or plant's organization. Consequently a sudden change of sufficient magnitude to produce a new species would almost certainly be eliminated by selection even if it arose.

Despite the difficulty in believing that species can arise at one step, except as the result of polyploidy (discussed below), Goldschmidt[30] has recently raised the issue again. He argues that the characters distinguishing species are of a different nature from those separating subspecies. There are, however, many borderline cases in which it is difficult or impossible to decide whether a form should be given specific or subspecific rank. This shows that there is no real distinction between the differences which separate species and those which separate subspecies, except with respect to their magnitude. Goldschmidt overcomes this difficulty of borderline examples by merely stating that they are subspecies, on the grounds that hybridization is sometimes possible, and by not recognizing specific difference unless he can find a considerable difference. This procedure unfortunately begs the whole question.

As an example of the absence of qualitative differences between characters determining species and subspecies we can take some members of the *glaucus* group of Swallowtail butterflies which inhabit North America. *Papilio glaucus* itself is found in the eastern United States as far west as the Mississippi basin, and it extends northwards into Canada and from there westwards to Alaska. The various geographical forms (subspecies) of this species differ in (i) the exact colour of the yellow on the wings, some being much more orange than others; (ii) the extent of the black stripes and bands on the wings; (iii) the amount of blue on the hind wing; (iv) the amount of red in a row of spots on the hind wing near the margin on the underside; (v) the distinctness of a row of yellow spots on the underside of the fore wing near its margin. These are usually separate and seldom tend to join up to form a yellow band; and (vi) in the south-eastern United States there is found a distinct black form which is a mimic of the Swallowtail *Battus philenor*.

In the United States, west of the Mississippi basin and extending as far north as Canada, there is an area inhabited by three species which are often found flying together in the same place at the same time, but do not hybridize and therefore are good species. These are *Papilio rutulus* Lucas, *P. multicaudatus* Kirby and *P. eurymedon* Bois. That they are distinct species is demonstrated by the fact that they can all be found flying together, and that they do not hybridize, thus allowing them to remain distinct. *Papilio rutulus* is very like the yellow form of *P. glaucus* but has little or no red on the hind wings, and the yellow spots on the underside of the fore wings are united to form a distinct band. It also differs in other small details of coloration, amongst other things tending to have much less blue than the southern form of *P. glaucus*. The yellow eye-like markings on the thorax of the larva also differ from those of *glaucus*, and it feeds on aspen and willow, which are not the normal food plants of *glaucus*. Nevertheless, the distinction between the two species is sufficiently tenuous for it to be almost impossible to identify some specimens with certainty.

Papilio eurymedon differs from both species in that the yellow on the wings of the female is far paler, and in the male it is so reduced as to be almost white. Furthermore, the black markings on the wings are more extensive. The blue on the hind wings is frequently much reduced, compared with *P. glaucus* the red on the underside of the hind wings is also less in evidence, and is often absent, and the yellow-white spots on the fore wing form a band. The larva differs in the yellow eye-spot marking on the thorax.

Papilio multicaudatus is larger and more orange than the other three species (with the possible exception of the Florida race of *P. glaucus* which is particularly large and particularly orange in colour). It also differs from the other three in that the margin of the hind wing is extended to form three tails instead of the usual one. With respect to the other characters, the blue tends to be reduced, the red absent and the yellow spots on the fore wings normally form a band. The range of food plants of *P. eurymedon*, *P. multicaudatus*, *P. rutulus* and *P. glaucus* differ, but all of them except *P. eurymedon* has at least one in common with *P. glaucus*.

It might be argued that *Papilio glaucus* and *P. rutulus* are only subspecies, for hybrids can be obtained *in captivity* by holding the male and female and manipulating the genitalia until the butterflies copulate.[13] Moreover, the hybrids are fertile, at least when backcrossed to *P. glaucus*. However, this argument on fertility is not valid. Similar hybrids have been obtained between *P. glaucus* and *P. eurymedon*, and these also are fertile when backcrossed to *P. glaucus*; therefore if one argues from fertility, one would have to maintain that all three species are really subspecies. But two of them, *P. rutulus* and *P. eurymedon*, fly together without hybridizing, thus showing that they are specifically distinct.

From the foregoing account it will be seen that the four species differ in the same characters as those distinguishing the subspecies of *Papilio glaucus*, but to a more marked degree: (i) in the exact shade of the yellow on the wings; (ii) the extent of black on the wing; (iii) the amount of blue on the hind wing; (iv) the amount of red on the underside of the hind wing; (v) the discreteness of the yellow spots on the underside of the fore wing; and (vi) the presence or absence of a black form. Of course, not all the species differ by all these characters; for example none but *P. glaucus* has a black form. It is true that the eye-spots on the larvae are not the same in the four species, but even in *P. glaucus* this character is variable. Moreover, although the range of food plants of the larvae is different, at least some of the species have one in common.

Thus we see that the differences between the four butterflies are of the same type as those which distinguish the various forms of *Papilio glaucus*, and are not qualitatively different. It is also obvious that distinct species or their hybrids are not invariably sterile if they can be persuaded to copulate, even if they will not do so in the field (exactly the same situation is found in other animals). We can therefore conclude that Goldschmidt's contention that species are qualitatively different, whereas subspecies are quantitatively different, is not valid. Moreover, genetic studies which are still in progress show that the differences between these Papilios are controlled by many genes, and not just one.[12] This finding is in absolute agreement with similar breeding experiments using species in other groups of organisms.

The hypothesis that species must arise by single major mutations can, therefore, be seen to be false. The hypothesis can also be rejected on other grounds. No mutation converting one species into another has ever been observed (excluding polyploidy) and, even should one arise, it would have very little chance of surviving (p. 183). Furthermore, because species do not usually hybridize, and because mutants first appear as heterozygotes, it would require at least one male and one female (in bisexually reproducing forms) to be present at the same time at the same place for the mutant to be perpetuated, even granted that such a form could be viable. This would only be likely to occur in organisms which can also reproduce by asexual means, and by this method accumulate a number of the mutants in the population. As is noted below, the production of species by autopolyploidy tends to be confined to such forms.

Thus having rejected the hypothesis that, in organisms which are necessarily cross-fertilizing, species normally arise by single steps, we can reduce the problem of their formation to that of the evolution by a series of small stages of isolating mechanisms which reduce or prevent gene exchange. In a population of cross-fertilizing organisms, any gene which tends to produce sterility between an individual carrying the gene and one not possessing it will tend to be at a disadvantage and be eliminated. This follows from the fact that in the early stages of its spread, an individual carrying the gene will nearly always fertilize, or be fertilized by, one not carrying it. Such unions will tend to be sterile and, therefore, the gene causing the sterility will be at a disadvantage and will be eliminated. If it becomes sufficiently common because it has some other effect which is beneficial, the disadvantage of the sterility mechanism will be reduced, and may even be converted into an advantage. The overall advantage of the gene will thus be increased and it will replace its allelomorph. Consequently two forms which are sterile when crossed are unlikely to be evolved from a single form in *one* population as the result of the accumulation of genes causing sterility.

Isolating mechanisms might also be evolved by the accumulation of allelomorphs reducing random mating in such a way that similar types tended to fertilize each other. However, within one population, such a gene, if it arose, would be at a disadvantage simply because genotypes like itself would be scarce, and it would therefore have little chance of finding a mate, and so such genes would lead to reduced fertility in individuals carrying them. Furthermore, even if such a gene did manage to spread for other reasons, the situation would not be stable, for its advantage would increase with increasing frequency, so that both allelomorphs would not be maintained. Thus the population would not split into two non-interbreeding groups. (For a possible exception to this rule see p. 173.)

If we argue that the absence of genetic exchange between species must result from the gradual accumulation of genes reducing genetic exchange between them, then we must conclude that for a species to evolve into two distinct forms it must be divided up into two or more separate populations. The exchange of genes between these must be much reduced or absent. In other words, with certain exceptions (*see below*) two species or distinct races cannot be formed from one in a single population. Before there can be speciation the population must be divided into non-interbreeding groups by spatial barriers, or by some other non-genetic impediment to the free exchange of genes.

When a species is divided up into separate populations, each will accumulate different genes as the result of natural selection, because the environments in which they live will be different, as will the particular range of genetic variability available to selection, as explained in Chapter VII. Consequently the two populations will evolve independently of one another by the accumulation of many small genetic differences. In time the individuals from the two areas may become so different that even if conditions change, and they meet, no hybridization occurs. If this stage is reached two new species will have been evolved. If they have not diverged thus far (that is to say, they are still subspecies), and they again come to occupy the same area (their ranges meet and overlap), there will be free interbreeding. This will result in an increase in variability in the population by the segregation of genes, and the distinctness of the two forms will disappear. This process, producing increased variability on which natural selection acts, has been called *introgressive hybridization*, and is of particular importance in plants.

Two diverging populations may, however, have reached a stage at which they are sufficiently distinct for the hybrids between them to be at a great disadvantage, but not sufficiently distinct for no hybridization to occur. In these circumstances the two forms will probably evolve into full species, as pointed out by Dobzhansky.[18] He argued that if the hybrids are at a disadvantage and tend to be eliminated, any individuals of the two forms which hybridized would leave less offspring to future generations than those that did not. But the likelihood of crossing is almost certainly affected by genetic components. Consequently those genotypes which reduce the probability of fertilization by the other form will be at a great advantage. These will increase, thus reducing and finally eliminating cross-matings.

K. F. Koopman[47] has confirmed that genes reducing hybridization can be selected in this way. He worked with the fruit-flies *Drosophila pseudoobscura* and *D. persimilis*, which only very rarely form hybrids in the wild, but do so readily in the laboratory. He kept these two species together in four cages and removed all the hybrid offspring in each generation. At the beginning of his experiments between 22% and 50% of hybrids (it varied from cage to cage) were formed in each generation, but after only six generations of this selection the number of hybrids was reduced to 5% or below. In later generations the proportion of hybrids dropped to only 1% in some populations. This experiment demonstrates rather elegantly the evolution of an isolating mechanism. Similar results have been obtained by Bruce Wallace and independently by G. R. Knight, A. Robertson and C. H. Waddington[46] within one species. They kept stocks of *D. melanogaster* differing by recessive genes, and removed the heterozygotes in each generation. This was quite a simple matter, for naturally these were wild type in appearance. After a number of generations the number of heterozygotes formed was materially reduced, again showing the evolution of an isolating mechanism.

It should be noted here that although the argument that isolating mechanisms evolve slowly, and that a species must be divided before it can evolve into two new ones, is generally true of animals, it is less true of plants. In some of them there is no logical reason why a mutant or an autopolyploid reducing random mating, or causing sterility in certain crosses, should not spread, for if the plant is able to build up a population of these genotypes by asexual means there will be no difficulty in the initial stages of their increase.

Unlike the situation in animals, polyploidy in plants is an important factor in the evolution of species, genera and higher taxonomic groups. That this is so depends not only on the fact that many plants have means of vegetative reproduction, but also on the fact that doubling the chromosome number does not usually have a profound effect on the morphological and physiological characters of an organism, although these are altered to some extent. As has been previously explained, any very large change is almost certain to be deleterious to a highly organized system. Consequently, if polyploidy produced such changes, it would be rare in plants as well as in animals.

Hybrids between species are often infertile although viable, as for example in the well-known case of the mule which results from a cross between an ass and a horse. When the sterility results from a failure of the chromosomes to pair at meiosis because they are very different, doubling their number to form an allopolyploid can restore the situation because each has then an identical partner with which it can unite (p. 56). Moreover, even when pairing in the hybrid is regular and sterility results from other causes (probably the presence of one set, not two, of genes from each parent, giving an unbalanced gene-complex), fertility may be restored in the polyploid. Allopolyploids (p. 56), formed by chromosome doubling in a sterile hybrid, are usually 'infertile' with the parent species (produce sterile triploids) and the hybrid. Consequently, new species can be formed in this way. An 'artificial' species of Primula, *Primula kewensis* Hort. (haploid number 18), was produced in England by a doubling of the chromosome number in a sterile hybrid between *P. verticillata* Forsk. and *P. floribunda* Wall. (both with a haploid number of 9). Although a new species can be formed at once in plants by allopolyploidy there may still be cross-pollination between it and one or other of its parent species. The hybrids will be triploids, sterile and therefore at a great disadvantage. Consequently there may be strong selection for mechanisms reducing cross-pollination between the allopolyploid and its parent species (p. 188).

The formation of new species in nature by allopolyploidy has been demonstrated by laboratory investigations. Thus A. Müntzing[54] suspected that the wild tetraploid Hemp-nettle *Galeopsis tetrahit* L. was an allopolyploid formed from a hybrid between *G. pubescens* Less and *G. speciosa* Mill. Moreover, by obtaining the hybrid between them and then, among other things, doubling the chromosome number using a series of crosses he was able to produce a plant not only like *G. tetrahit*, but also fully fertile with it. Thus he resynthesized a species which had previously been formed in nature. An equally interesting example is found in the cord grass Spartina. An American species *Spartina alterniflora* (Lois) established itself on the south coast of England and hybridized with the European species *S. maritima* (Fernald.) The hybrid produced an allopolyploid which is fully fertile and apparently better adapted than either parent species, so that it has spread rapidly along the sea coasts. Had the chromosome doubling drastically affected the morphology or physiology of the new form, other than with respect to fertility, it is unlikely that it could have persisted. From the available data on both autopolyploidy and allopolyploidy we are led to the conclusion that, although the sterility barrier between species may arise at one step in plants, and but rarely in animals, a marked morphological or physiological difference between species must in most cases be evolved gradually as the result of a series of small steps.

The results of recent research have shown that in the higher animals species evolve gradually, and that for one to split into two there has to be some form of spatial isolation, at least in the early stages of the process. The same is true of some plants, particularly annuals which reproduce sexually. However, in others, notably perennials with efficient means of asexual reproduction, a new species can arise at one step as the result of doubling the chromosome number. Such species are usually very like their diploid ancestors and only diverge from them as the result of the slow accumulation of small differences, the sudden appearance of a markedly distinct species by polyploidy, or by other means, being very unlikely.

Both Darwin and Wallace were correct in believing that most

of the differences between species are evolved gradually and over a long period of time. They could not have been expected to know that the hybrid sterility between species can, on occasion, occur by one step, for they knew nothing about chromosomes or Mendelian genetics.

CONCLUSIONS ON EVOLUTION AND SELECTION

Adaptation

HAVING considered some of the ways in which natural selection acts, we must enquire whether it can explain the evolutionary changes shown by the fossil record, and indicated by the diversity of living things found on the earth at the present time. This can be done best by considering only those features of both which are not obviously explicable on the hypothesis of evolution by means of natural selection.

The difficulty of accounting for the evolution of such complex organs as the eye or the electric organ of fish has already been discussed. The evolution of a very specialized life cycle can be illustrated from the European Large Blue butterfly, *Maculinea arion* (L.), although even more extraordinary examples could have been picked. The butterfly lays its eggs on the flower heads of Wild Thyme and the young larvae feed on these for the first few weeks of their life. They are great cannibals and the larger larvae eat the smaller ones whenever they get the chance to do so. On the seventh abdominal segment of the larva there is a small gland which is very sensitive to touch. While the larvae are still feeding on Thyme this gland becomes functional and, if stimulated, it produces a drop of sweet fluid. Ants attend these larvae and 'milk' them for this secretion in much the same way that they 'farm' aphids for the sweet liquid that these produce. After about three weeks the caterpillar leaves the Thyme, never to return, and wanders on the ground in an apparently aimless manner. If a foraging ant meets the larva, it will milk it repeatedly. After a time the caterpillar expands its body just behind the head and in front of the gland. The ant then seizes it in its jaws by the expansion and carries it off into its nest, which may be some distance away. For the rest of the summer the larva feeds on the young of the ant. In the winter it hibernates but starts feeding again in the spring.

Later it pupates in the nest and, after about three weeks, the butterfly emerges from the pupa and crawls through the passages of the nest to the outside world, where it has not been since it was carried into the nest as a larva. The butterfly then expands its wings, a process which is delayed until it is clear of the nest.

At first sight it might seem difficult to understand how such a life history could have been evolved gradually. However, the relatives of this species show intermediate degrees of adaptation. It is obviously advantageous for a defenceless caterpillar to have ants in attendance, for they are powerful insects, and capable of protecting it from would-be predators. In some species, for example the Large Copper *Lycaena dispar* (Haworth), the sweet secretion is produced from scattered gland cells, and not from a specialized region. The production of such a fluid could be evolved from cells originally evolved to get rid of waste products, or to coat the larva with wax in order to keep down water loss by evaporation. A sweet secretion could be advantageous in directing the ant's attention away from eating the larva itself if the fluid is more attractive, and this could be changed later into a means of getting protection from other predators or from parasites. Now a concentrated drop is likely to be more attractive than a thin film, so that individuals in which the gland cells were not scattered but clumped would be at an advantage and, by selection for this attribute, a specialized gland could be evolved.

The interrelationship between ants and the caterpillars of some species, which are not taken into the ants' nests, such as the Brown Argus, *Aricia agestis* (Schiff.), has been developed so far that the larva does not survive well even in captivity, unless attended by ants. The reason for this is that fungi grow in the drop of fluid from the gland if it is not regularly removed. Species which have evolved this close association are often moved by the ants to suitable plants near the nest just in the same way as ants move aphids. This is clearly of advantage to the ants, for they then do not have to go so far to milk their 'stock'. Now aphids are not cannibalistic, as are the larvae of many of the Blues. Consequently an aphid which was carried into the ant's nest would be likely to die, but the larva of some species of Blue might well live because it could feed on the young of the ants. Moreover, any individual that did so would be protected from predators, parasites, rain, cold and other factors which take a high toll of larvae and pupae. Thus there are a series of steps, each advantageous in some respect, either to the larvae or to the ants, or both, which would enable a complex life history of this type to be evolved.

The example illustrates another aspect of the result of long-continued selection for a specialized existence. Although the species becomes very well adapted to its particular mode of life, it may thereby become extinct, for it becomes dependent on some particular aspect of the environment, in this example ants, and any sudden change may be fatal to it. Thus selection often in the long run leads to extinction, and it is those species which have not specialized which tend to survive to give rise to new forms. We would, therefore, expect to find rather few such specialized foerm in any one group at any one time, and, though they may be very successful temporarily, they will, in the end, become extinct. In view in the mode of existence of the Large Blue, it is not surprising that of England, where it is at the edge of its range, it has disappeared from many of its old haunts in the last one hundred years.

THE FOSSIL RECORD

Evolution by sudden transition or saltation

It has frequently been pointed out that species and genera often appear suddenly in the fossil record, and this has been interpreted as meaning that they arose suddenly by 'mutation' or special creation. The record is, however, by no means complete, and the remains of most of the species which have existed have probably never been preserved. To take an example, a group of fish, the coelacanths, are known as fossils. They first appeared some 270 million years ago, and disappeared from the record some 60 million years ago. However, despite the absence of any fossils during the last 60 million years, members of the group still survive. One, called *Latimeria*, was caught just before the Second World War in the sea off Africa, and several have been caught since. Thus the absence of remains in the fossil record for 60 million years did not mean that the fish did not exist.

Even if there is an almost complete record, a new form can suddenly appear without this meaning that it has suddenly been evolved. For example, in the evolution of the horse there were known from Europe and Asia a number of distinct groups which appeared one after the other as fossils, but each did not evolve to any large extent, died out, and was replaced by a new type. This suggested that the new forms arose at one step. However, in the last sixty or seventy years, many fossil horses have been found in North America, and it has been shown that their evolution was gradual. The sudden appearance of successive waves of new types in Europe resulted from successive invasions from North America.[63]

Long-continued trends in evolution

In contrast, many fossil lineages show a gradual change in the form of an organism over a period of many millions of years. This observation, which is almost the exact reverse of the previous one, has also been adjudged fatal to the theory of evolution by natural selection. The evolution of the horse is again often quoted as showing such a steady change. However, as G. G. Simpson[63] has pointed out, this is not a true interpretation of the facts. The earliest known members of the horse 'family' were about the size of a medium-sized dog (10–20 in. high), and had four toes on the fore and three on the hind feet, each terminating in a separate hoof. The teeth were low-crowned and simple in structure, being suitable for grinding the food. The front teeth were modified for plucking or picking up leaves. Simpson has shown that none of the special features of the modern horse were evolved at a steady rate in a constant direction, but that the speed and the direction in which they evolved varied from time to time, as would be expected if natural selection were causing the change. The line which eventually gave rise to the horses shows a reduction in the number of toes until only one is left on each foot. At the same time the size of the animals increased as did the size of the teeth, but, at one stage, a new form of tooth was selected for. This was high-crowned with an intricate pattern of crests and ridges which made it possible for the animals to change their diet from browsing on leaves to grazing on grass. These are particularly tough plants which would soon wear down the low-crowned teeth of the browsing horses. This new type of tooth was evolved at a time when grassland was becoming commoner, and the broad-leaved forests of earlier times were receding.

Life in open country favoured swifter, larger animals which were more easily able to see and avoid their predators. Consequently, selection favoured an increase in body size and a reduction in the number of toes, for this gave greater speed. At the same time the teeth also became larger and more durable, a necessary change to allow for the greater volume of food taken by a larger animal. Thus we see that the evolution of the modern horse resulted from long-continued selection for increased size and speed, together with selection for those forms which could utilize the most abundant sources of food; first leaves, and later, with a change in climate, grass. During this time other types were also evolved, but these were in the end less successful and became extinct.

The evolution of horses illustrates the point that selection may have to operate in the same direction for an immense period of time before the organism attains optimum proportions. Thus in the horse each increment in size necessitated changes in the strength of the limbs and toes, as well as in the structure of the teeth, and many other less obvious alterations. Although a very small increase in size might be favoured at any one time, too large a change would be deleterious because the other characters would then be out of proportion. But a very small increase in size would only give a very small advantage with respect to speed, strength and field of vision. Consequently, although selection in this direction was probably long sustained, its net magnitude was very small and, therefore, progress was very slow (*see also* p. 111).

Momentum in evolution

Long-continued evolutionary trends are often sustained up to the moment that the species becomes extinct. This has been interpreted as showing that evolution can gain a 'momentum' which causes characters to overshoot their optimum expression, and so, having become deleterious, cause the extinction of the species. The most widely quoted example of this is the extinct giant Irish Deer, *Megaceros*. The ancestors of this deer were small and had quite unexceptional antler proportions. During the evolution of the group, however, the animals tended to get larger and larger, as did the horses, but the size of the antlers in comparison to that of the deer became even greater, so much so that many workers have maintained that these huge structures must have been very disadvantageous, and caused the extinction of the deer. However, not only is there no known mechanism which would give momentum to evolution, but there is no necessity for postulating that the relative proportions of antler to body size was not controlled by natural selection.

The parts of the body do not grow at the same rate as the body as a whole, so that the proportions alter with age. J. S. Huxley[40] has investigated the matter in considerable detail, and shown that this proposition holds in many groups throughout the animal kingdom. Thus it is a general feature of deer that the antlers of the male increase in size more rapidly than the body as a whole. Consequently, large stags have relatively larger antlers than small stags of the same species. This is true of the Irish Deer, so that the older, and therefore on average larger, animals have relatively larger antlers than the younger ones. But older animals have a shorter expectation of life and, therefore, will in general have less progeny in the remaining part of their life than younger animals living at the same time; that is to say, the older ones will have already had most of the offspring which they are going to have during their life. Thus any character which is deleterious or lethal in later life will not be subjected to very strong selection-pressure against it, for it will have very little effect on the number of progeny the animal leaves. If the character only manifests itself after the reproductive period is over, there will be no selection against it (excepting in social animals when there is co-operation between individuals who are members of the same family). Consequently, if in the Irish Deer the larger young stags with their large antlers were reproductively more successful than the smaller ones (which seems likely), there would be strong selection for increased size of body and antlers even if these attributes caused death later in life. There is no reason to postulate 'momentum' in evolution to account for the facts. It is much more probable that the Irish Deer became extinct because of a change in the vegetation in the area where it was living, and there is, in fact, some evidence for this view.

The pattern of evolutionary change

An examination of the vertebrate fossil record will show that when a new major group or class appears for the first time, there is usually a sudden burst of evolution. Many new forms appear and these give rise to the main types within the major group. Thereafter, although the members of the main types become specialized to various modes of life, entirely new types are rarely evolved from them. Finally, a stage sets in when there is very little in the way of major evolutionary change and many main types become extinct. This has been interpreted as meaning that these groups, like individuals, have a 'life history' consisting of 'youth', 'maturity' and 'old age', thus implying that the evolutionary history is predetermined by some unknown force, and is not directed by natural selection. However, a new major group usually depends on the appearance of some new 'invention' which enables the organism to exploit the environment in a new or more efficient way and consequently evolve yet other 'inventions'. Thus the acquisition of lungs allowed the fishes to invade the land and evolve into amphibians. Consequent upon this depended the evolution of an egg which could conserve water and so did not have to develop in a pool or damp place. This device allowed the amphibia to live in more arid places and evolve into reptiles. Wings and heat regulation have made it possible for the birds to utilize a far wider range of environments for breeding and gathering food. The evolution of mechanisms allowing heat regulation, together with the development of the young within the mother and the feeding of the young with milk, allowed mammals to live, feed, remain active and breed in far more severe conditions than almost any reptile could. Thus the evolution of lungs made habitable a whole new environment which resulted in the evolution of yet other new inventions.

The appearance of such inventions opens up new ways of life, so that one would expect a burst of evolution within the major group as these become exploited. However, very shortly there will be no new modes of life open, for all that can be will have been utilized by one or other of the forms which have been evolved. Consequently, changes will be confined to increased specialization and adaptation in response to environmental changes both climatic and those due to other organisms evolving. At a later stage lines which have become over-specialized will tend to die out as conditions alter. Only a new invention is likely to give rise to a new major group and therefore to a new burst of evolution. Thus we see that change controlled by natural selection accounts adequately for the pattern of evolution found in the major groups of vertebrate animals.

Parallel and convergent evolution

In view of the fact that certain characteristics are advantageous in a large number of different environments, it is not surprising that similar trends, such as an increase in size, appear in many different lines. To take another example, as flight is a very effective way of avoiding earthbound predators, it is not surprising that gliding flight has been evolved independently more than once in fish, amphibia, reptiles and mammals. Moreover, true flight has also been evolved independently in reptiles (the pterodactyls), birds and mammals (the bats), as well as in the insects. This method of locomotion, once evolved, opens up, among other things, new food sources and modes of dispersal. With regard to selection for the efficient dispersal of young, animals such as spiders and some insect larvae have evolved a method of parachuting by producing a silken thread from specialized glands, and remaining attached to it while it is blown along by the wind. Young spiders are known to have drifted as much as four hundred miles in this way. Plants also have evolved a very large number of parachute and wing-like devices to facilitate seed dispersal by wind. When special conditions recur, often the same specializations are evolved to exploit them. 'Horses' for example have been produced twice, once giving the true horses (p. 195) and once quite separately in South America, from archaic mammals, producing forms incredibly similar (*Diadiaphorus* and *Thoatherium*). Mimicry is another example, although because of the circumstances the resemblance is more superficial (Fig. 8).

Conclusion

Genetical and ecological investigations, coupled with field studies, have revealed many of the agents which control adaptation and speciation. Those underlying the evolution of the higher taxonomic groups (Families, Orders, Classes, etc.), cannot be determined so readily. However, data taken from several sources, particularly anatomy, embryology and palaeontology, suggest that the agents are in no way different from those controlling speciation. All that is required for the evolution of the higher groups is the long continued action of the same processes which have been shown to be occurring within wild populations at the present time.

We can conclude that not only has natural selection occurred, but that it is competent to account for the facts of adaptation and evolution, as we know them. It is, moreover, the only hypothesis which will explain them adequately.

REFERENCES

1. ALLISON, A. C. *Ann. Hum. Genet.* (1956), *21*, 67–89.
2. BATES, H. W. *Trans. Linn. Soc. Lond.* (1862), *23*, 495–566.
3. BIRCH, L. C. *Evolution* (1955), *9*, 389–99.
4. BLEST, A. D. *Behaviour* (1957), *11*, 209–56, 257–309.
5. BUMPUS, H. C. *Biol. Lectures Mar. Biol. Lab. Woods Hole Lect.* (1898), *11*, 209–26.
6. CAIN, A. J. *Animal Species and their Evolution* (Hutchinson's, London (1954)).
7. CAIN, A. J., and SHEPPARD, P. M. *Genetics* (1954), *39*, 89–116.
8. CAIN, A. J., and SHEPPARD, P. M. *J. Genet.* (1957), *55*, 195–9.
9. CARPENTER, G. D. H., and FORD, E. B. *Mimicry* (Methuen, London (1933)).
10. CASPARI, E. *Amer. Nat.* (1950), *84*, 367–80.
11. CESNOLA, A. P. DI. *Biometrika* (1907), *5*, 387–99.
12. CLARKE, C. A., and SHEPPARD, P. M. *Evolution* (1955), *9*, 182–201.
13. CLARKE, C. A., and SHEPPARD, P. M. *Lepid. News* (1955), *9*, 46–8.
14. CLARKE, C. A., MCCONNELL, R. B., and SHEPPARD, P. M. *Lancet* (1957), *272*, 212.
15. COTT, H. B. *Adaptive coloration in animals* (Methuen, London (1940)).
16. CROSBY, J. L. *Evolution* (1949), *3*, 212–30.
17. DIVER, C. *Nature, Lond.* (1929), *124*, 183.
18. DOBZHANSKY, TH. *Genetics and the origin of species*, 3rd ed. (Columbia Univ. Press, New York (1951)).
19. DOWDESWELL, W. H., FORD, E. B., and MCWHIRTER, K. G. *Heredity* (1957), *11*, 51–65.
20. FISHER, R. A. *The genetical theory of natural selection* (O.U.P., London (1930)).
21. FISHER, R. A. *Ann. Eugen. Lond.* (1939), *9*, 109–22.
22. FISHER, R. A., and HOLT, S. B. *Ann. Eugen. Lond.* (1944), *12*, 102–20.
23. FORD, E. B. *Amer. Nat.* (1930), *64*, 560–6.
24. FORD, E. B. *Biol. Rev.* (1937), *12*, 461–503.
25. FORD, E. B. *Ann. Eugen. Lond.* (1940), *10*, 227–52.
26. FORD, E. B. *Advanc. Genet.* (1953), *5*, 43–87.

27. FORD, E. B. *Heredity* (1955), *9*, 255–64.
28. FORD, E. B. *Genetics for medical students* (Methuen, London (1956)).
29. FRACCARO, M. *Ann. Hum. Genet.* (1955–56), *20*, 282–97.
30. GOLDSCHMIDT, R. B. *The material basis of evolution* (Yale Univ. Press, New Haven (1940)).
31. GOLDSCHMIDT, R. B. *Quart. Rev. Biol.* (1945), *20*, 147–64, 205–30.
32. HALDANE, J. B. S. *J. Genet.* (1939), *37*, 365–74.
33. HALDANE, J. B. S. *Proc. Roy. Soc. B.* (1948), *135*, 147–70.
34. HALDANE, J. B. S. *New Biol.* (1953), *15*, 9–24.
35. HALDANE, J. B. S. *Proc. Roy. Soc. B.* (1955), *144*, 217–20.
36. HALDANE, J. B. S. *Proc. Roy. Soc. B.* (1956), *145*, 306–8.
37. HARLAND, S. C. *Heredity* (1947), *1*, 121–5.
38. HARLAND, S. C., and ATTECK, O. M. *J. Genet.* (1941), *42*, 21–47.
39. HARRISON, B. J., and MATHER, K. *Heredity* (1950), *4*, 295–312.
40. HUXLEY, J. S. *Problems of relative growth* (Methuen, London (1932)).
41. HUXLEY, J. S. *Evolution: the modern synthesis* (Allen & Unwin, London (1942)).
42. HUXLEY, J. S. *Heredity* (1955), *9*, 1–52.
43. KARN, M. N., and PENROSE, L. S. *Ann. Eugen. Lond.* (1951–52), *16*, 147–64.
44. KETTLEWELL, H. B. D. *Nature, Lond.* (1955), *175*, 943–4.
45. KETTLEWELL, H. B. D. *Heredity* (1956), *10*, 287–301.
46. KNIGHT, G. R., ROBERTSON, A., and WADDINGTON, C. H. *Evolution* (1956), *10*, 14–22.
47. KOOPMAN, K. F. *Evolution* (1950), *4*, 135–48.
48. LACK, D. *The natural regulation of animal numbers* (Clarendon Press, Oxford (1954)).
49. LISSMANN, H. W. *Nature Lond.* (1951), *167*, 201.
50. LORENZ, K. Z. *King Solomon's Ring* (Methuen, London (1952)).
51. MATHER, K., and HARRISON, B. J. *Heredity* (1949), *3*, 1–52, 131–62.
52. MAYR, E. In *Evolution as a process* (Allen & Unwin, London (1954)).
53. MOSTLER, G. *Z. Morph. Ökol. Tiere.* (1934–35), *29*, 381–454.
54. MÜNTZING, A. *Hereditas Lund.* (1932), *16*, 105–54.
55. ROBSON, G. C., and RICHARDS, O. W. *The variation of animals in nature* (Longmans, London (1936)).
56. RUITER, L. DE. *Countershading in caterpillars* (Thesis, Univ. Groningen (1955)).
57. SCHNETTER, M., and SEDLMAIR, H. *Naturwissenschaften* (1953), *40*, 515–16.
58. SCHWAB, J. J. *Genetics* (1940), *25*, 157–77.

59. SHEPPARD, P. M. *Heredity* (1951), *5*, 125–34.
60. SHEPPARD, P. M. *Heredity* (1952), *6*, 239–41.
61. SHEPPARD, P. M. *Symp. Soc. Exp. Biol.* (1953), *7*, 274–89.
62. SHEPPARD, P. M. *Amer. Nat.* (1953), *87*, 283–94.
63. SIMPSON, G. G. *Tempo and mode in evolution* (Columbia Univ. Press, New York (1944)).
64. SOUTHERN, H. N. In *Evolution as a process* (Allen & Unwin, London (1954)).
65. STEBBINS, G. L., and FERLAN, L. *Evolution* (1956), *10*, 32–46.
66. SWYNNERTON, C. F. M. *3rd Int. Ent. Cong. Zürich* (1926), 2, 478–506.
67. TIMOFÉEFF-RESSOVSKY, N. W. *Biol. Zbl.* (1940), *60*, 130–7.
68. TINBERGEN, N. *The study of instinct* (Clarendon Press, Oxford (1951)).
69. TRIMEN, R. *Trans. Linn. Soc. Lond.* (1869), *26*, 497–521.
70. WADDINGTON, C. H. *Evolution* (1953), *7*, 118–26.
71. WADDINGTON, C. H. *Evolution* (1956), *10*, 1–13.
72. WELDON, W. F. R. *Biometrika* (1901), *1*, 109–24.
73. WHITE, M. J. D. *Animal cytology and evolution*, 2nd ed. (Cambridge Univ. Press (1954)).
74. WINDECKER, W. *Z. Morph. Ökol. Tiere.* (1939), *35*, 84–138.
75. WRIGHT, S. *Amer. Nat.* (1934), *68*, 24–53.
76. WRIGHT, S. *Evolution* (1948), *2*, 279–94.

INDEX

Bold type indicates page on which scientific term is defined.

204